Partnering for Success

Thomas R. Warne

ASCE
PRESS

Published by
ASCE Press
American Society of Civil Engineers
345 East 47th Street
New York, New York 10017-2398

ABSTRACT

Partnering as a method of doing business is spreading throughout the construction industry. With everyone fed up with the litigious nature of the industry, partnering represents an opportunity for owners, designers, contractors, subcontractors, and suppliers to maximize their individual abilities in a synergistic arena. In his book, *Partnering for Success,* Tom Warne explores the underlying principles of partnering and describes how one can develop successful partnering relationships. Mr. Warne discusses such issues as: 1) Partnering workshop development; 2) charter development; 3) team expansion; 4) value engineering; and 5) joint evaluation. He also provides samples of actual partnering documents, a sample partner charter, and a sample joint evaluation form. As presented in Tom Warne's book, partnering will serve as the catalyst for many improvements in the process of delivering products and services that have been engineered and constructed.

Library of Congress Cataloging-in-Publication Data

Warne, Thomas R.
    Partnering for success/Thomas R. Warne
      p.cm.
    Includes index.
    ISBN 0-87262-976-7
      1. Contractors operations—Management. 2. Industrial project management.
    3. Partnership. 4. Negotiation in business. I. Title.
    TA210.W37 1994   94-7661
    692'.8—dc20   CIP

Acknowledgments

First and foremost, I thank my wife Renae and our children for their support in my many endeavors. Their encouragement and understanding have been unfailing during these exciting times.

Larry Bonine first introduced public sector partnering in 1988 to the U.S. Army Corps of Engineers as the Commander of the Mobile (Ala.) District. From there the concept has spread throughout the country and is used in nearly every state. Among the pioneers of partnering, there are some who have gone beyond the mark in their dedication towards its establishment. Chuck Cowan and the partnering concept are synonymous. Other public sector champions include Norm Anderson, Ron Williams and August Hardt. From the industry, Doug Pruit, Dick Lewis and Ted Haworth have contributed much through their efforts to establish this great concept within their own firms and throughout the industry.

No work such as this book would be possible without the efforts of many people. My wife Renae spent many hours editing and suggesting improvements that would make this book more valuable to the reader. Zoe Foundotos, Gary Bates and John Hribar were also very generous in offering assistance that contributed to the quality of this effort.

To each I give thanks.

Foreword

I first became aware of Tom Warne through his reputation as a proponent of partnering and as a talented and inspiring speaker. I first spoke with him when I had been named as the Director of the Arizona Department of Transportation (ADOT) and, he as the Deputy, had phoned to congratulate me. I finally met him face to face on March 1, 1993, the day I started work at ADOT.

To appreciate Tom Warne's commitment and skill in the execution of a world-class partnering program, you have to spend some time around our office, on a job site, and in a regular partnering meeting with our industry suppliers and contractors. Somehow the physical atmosphere is different—more highly charged and exciting. People talk to each other and there is an air of trust and mutual respect that is more apparent than any other place I have worked. There is an expectation that something positive will happen. The total buy-in by and commitment of our industry partners has been the key to Arizona's success.

Partnering is not difficult, makes ultimate sense and is spreading faster than a highly contagious virus throughout the United States. Tom Warne has been a key evangelist in the growth of this powerful movement. He has delivered the message to organizations in 18 states, published numerous articles and is genuinely recognized as the principal guru of partnering today.

I think the reason partnering has caught on so fast and with such enthusiasm has a lot to do with timing. First everyone is fed up with the litigious nature of our industry. I have had countless construction company owners express their disgust with the legal fees they're paying. As a public sector owner, my peers and I just do not like the time and expense that litigation causes. The second reason is that the time was right has to do with the quality movement in the private sector. Partnering fits hand in glove with a quality initiative in any organization. Those students of quality know that decisions are best made by the process owner who is closest to the action. In partnering this is at the job site with empowered workers and project managers. Partnering workshops teach many of the skills characteristic of a Total Quality Management (TQM) environment and, simply stated, these techniques and skills work.

I am excited by all this. I am thrilled to be a part of it, and I'm delighted to have my name in this first great book on partnering with Tom Warne's. He wrote it, has the skill to make it work and best of all, he is my friend and we are on the same team. I encourage you to use this book, write in it, mark in it, and make it your "how to" comprehensive book on partnering. This is a wonderful time to be a part of America's turnaround in the engineering and construction industries.

Larry Bonine

Contents

Introduction

The construction industry in the United States historically has been known for its significant accomplishments. From towering skyscrapers that mark city centers across the nation to a world-class interstate freeway system that provides a vital link to our citizens, the construction industry can be proud of its past achievements. Today, however, the construction industry is not as renowned for its construction quality as it is for its confrontation, claims and litigation. The attributes of honor, integrity and pride in workmanship have been replaced with confrontation, claims, delayed projects, cost overruns and a host of other problems that render all parties less than fully effective. Countless hours and vast sums of money are invested by all parties in efforts to secure their positions in these non-productive activities.

These trends in claims and lawsuits have diverted funds that should have made owners more cost effective and contractors more profitable. Clear indications that something is wrong in the industry are all around us: project teams include full-time attorneys, some firms employ more attorneys than engineers, and "claims" seminars for both owners and contractors have proliferated—all reflections of major problems within the construction industry.

Another disturbing trend has been the tendency for projects to not be completed on time. The Arizona Department of Transportation found that in fiscal year 1991 over 25% of its projects were finished past the original completion date. Regardless of the reason for this delay, it impacted the profitability of contractors and the effectiveness of owners. Far more critical is the fact that this delay in project completion generally denied customers use of the final product.

Related to the confrontation between owners and contractors has been the win-lose approach prevalent on construction projects. The idea, "if I win you have to lose and vice versa," has been the axiom of most project personnel. Owner's representatives were trained to believe that it was appropriate to "jerk" the contractor around and to impose arbitrary or capricious orders on the contractor. These actions let the contractor know who was really in charge of the project. Work was needlessly rejected or held up by owner's representatives who thought they were properly administering a contract. They saw these activities as the means to properly look after the owner's interests in a project. Contractors, on the other hand, would react and endeavor to recoup losses in any number of ways, most of which were unknown to the owner. Ultimately, this win-lose approach was costly in both time and resources and resulted in a final lose-lose scenario.

The real tragedy of the last twenty years is the amount of resources that were expended on low-value or no-value activities. Low-value activities are generally in a form such as obstructing the other party's progress, withholding vital information, basing decisions on future legal maneuvering and so on. Whether you are an owner, designer, contractor or supplier, these costs have been substantial.

In addition, the years of confrontation have taken a huge personal toll on the lives of countless individuals in the industry. The hatred and negative feelings that have been characteristic of so many projects have opened deep wounds of cynicism and distrust. People were heard to say that their work "just wasn't fun any more."

In the late 1980s and early 1990s, some visionary leaders in the industry concluded that there had to be a better way. They believed that the redirection of all these negative, non-productive efforts would ultimately result in positive, tangible benefits. If owners and contractors could work together in a synergistic environment, then certainly everyone would benefit. Hence the concept of partnering was born and the construction industry was forever changed.

Total Quality Management and Partnering
In recent years there has been much discussion and activity in the United States regarding the concept of Total Quality Management (TQM). TQM has been the response of American business and industry to growing foreign competition and trends in reduced profitability. It focuses on customer service, continuous improvement and represents a fundamental shift in business and management philosophy for those who embrace it. TQM provides an opportunity for public and private entities to examine their basic culture and business practice and eliminate unnecessary work and rework. The resulting organization is more profitable and ultimately provides a higher level of customer service.

In any business you have both suppliers and customers. Partnering is nothing more than a concept that addresses both the supplier and customer components of TQM. In the partnering relationship, the design firm is a supplier of plans and specifications to the owner. However, the designer is also a customer because it must first receive a "project assessment" or "design concept report" and other information from the owner in order to fulfill its specific role on the project.

This combination of supplier/customer roles is not unique to partnering but does have an influence on the process. As organizations examine their many

2

relationships and processes they find that significant time and resources are invested with suppliers. The quality of supplier relationships and the effectiveness of their business process impacts the products and service customers receive. In examining the significance of this area of its business ADOT found it spent almost 70 cents of every dollar in supplier relationships. With such a large share of its resources focused in this area of its business it is clear why partnering became ADOT's first, and continues to be its most mature component of its Quality and Productivity Initiative (QPI).

Partnering recognizes that the construction project is nothing more than a compilation of many processes and the efforts of numerous customers and suppliers. By applying the principles of TQM through the partnering relationship, the owner and the contractor will find greater success and profitability in their respective roles.

Organizations that currently have active quality initiatives will find the use of partnering a natural extension of their many other efforts. Partnering will serve as the catalyst for many improvements in the process of delivering products and services that have been engineered and constructed.

Chapter 1

Principles of Partnering

The principles of partnering are simple and easily understood in their application. Partnering is nothing more than a return to the same basic values that have been fundamental to honorable societies for centuries. It should come as no surprise to the reader that the same principles that were valid years ago are still valid today as we interact with other members of our society.

Stakeholders

Partnering recognizes that there are many stakeholders on any given construction project. A stakeholder is defined as some individual or organization that has a vested interest in the successful completion of a project. A stakeholder may be an owner, a prime contractor, a design engineer or architect, a subcontractor, a supplier, a local community or business group, a governmental agency or any one of a host of others.

Each stakeholder has a role in the construction process of any given project. In addition each stakeholder has a specific definition of success that is unique to his or her role or perspective of the project. To the contractor, profit may be a significant part of its definition of success. To the subcontractor or supplier, success may be defined by the fact that no work failed to meet the specifications and timely payment was made. An owner may define success as the final product that can then be turned over to the user. An entity such as the Federal Highway Administration may include proper and thorough accounting procedures in its definition of success. The designer defines success as the proper implementation of the design that meets the owner's desires and expectations.

Partnering recognizes that, even though each stakeholder has its unique definition of success, these unique definitions are not mutually exclusive. The process recognizes that owners, contractors, suppliers and others can all be simultaneously successful on a project and that no one stakeholder's definition of success would preclude another's. Each partner or stakeholder can and should feel good about achievement on a construction project.

Common Goals

In addition to recognizing the many stakeholders on a project, there is also a realization that each of these stakeholders shares certain common goals. When each stakeholder examines its fundamental goals and objectives it finds these include some very basic issues. An owner would like to have a project built where work is completed to specification and there is no need to reject any work performed. The contractor would like to complete all of the contract items in such a way that none are rejected by the owner and the quality is sufficient that no rework is required. The designer would like to see that the design is properly and completely constructed.

Owners would also like projects completed on time so they can then be used for their intended purpose. The public or users, who are very important stakeholders, want the use of the completed work as soon as possible and with the least amount of inconvenience in the process. With few exceptions, owners and contractors would prefer to avoid claims or litigation incident to a construction project. In addition, all parties to a contract are interested in safety for both their employees and those who will come in contact with the work.

In the partnering process, the stakeholders gather together and compare their organizational goals to those of other team members. Ultimately they come to realize that they share common goals such as quality, timely completion, safety, no rework, and no claims. Once these common goals are recognized, the team members then must ask themselves the question: "If we share these common goals, then why can't we work together toward their achievement?" The answer is that the partnering team can and should work together toward these common goals.

Attributes of Partnering
The attributes of partnering are simple in concept and also in their application. Fundamental to the process is the creation of a high-trust culture. A high-trust culture is one where parties rise above the deceit, distrust, innuendoes and hidden agendas of the old construction process. Putting these negative characteristics aside, the partnering team engages in the more productive and open relationship of honesty, trust and synergy.

Establishing this high-trust culture will take time and effort on everyone's part. It took years of bad experiences to develop the very poor relationships that exist in the industry today. This being the case, it will take time and many positive experiences to overcome this "deficit-type" relationship.

While it would be nice to write a policy that required trust or perhaps pass legislation saying that "we all trust each other," this is simply not how it will be developed. Trust will come slowly and will develop little by little as individuals work together and demonstrate that they are worthy of trust. The issues that we begin to trust on will be simple at first and will bear little risk to the parties. Eventually the issues will become greater and the level of trust will grow. As parties come to realize that their counterparts are sincere, competent and trustworthy, the relationship will proceed toward this high-trust culture.

The high-trust culture allows parties to come to the table with their problems and issues with the assurance that they will not be taken advantage of or treated unfairly. It encourages team members to come forward and share in the resolution of problems rather than hiding them for fear that costly repercussions may occur. This culture requires that team members accept the mistakes and errors of others and assist them in finding solutions that are mutually agreeable to all of the team members. This is not to say that one or more parties can abrogate their contractual responsibility through the partnering process. On the contrary, each party is faithful to their respective role and takes responsibility for all actions related to the contract.

In the high-trust culture, organizations treat one another in a trustworthy manner. Too often specifications and contracts have been written in a way to deal with the small percentage of untrustworthy contractors or suppliers. Basically this process has punished the masses in an attempt to police the few. The high-trust culture creates an environment where individuals and organizations are treated as being fundamentally honest and possessing integrity.

Synergy is an important attribute of the partnering process. It is simply a matter of individuals working together and expending their energies in a team effort as opposed to their many individual efforts. Synergy is a natural outcome of partnering and contributes to the overall success of the project as a whole. It recognizes that an owner, contractor and designer working together can produce a higher quality product than each of the three working separately on the same product. In this case, one plus one plus one equals four or five or more: the multiplicative effects of the synergistic relationship take over.

Synergy is the ingredient that causes good things to happen on a construction project. Value engineering ideas come out of synergistic relationships. The synergistic team meets problems head on during the partnered project with solutions sought from a team perspective.

Top Management Commitment

As with any new and innovative program, effort is required to implement the idea and move it toward success. It should come as no surprise that top management commitment is essential to the success of the partnering process in any organization. This fact has been proven across the country in organizations that have implemented partnering. Regardless of the type of organization, the evidence is clear—success is absolutely dependent on the commitment at the executive level.

Top managers must not only say the right words, but also they must "walk the talk." They must be disciples of the partnering process and be unashamed to spread the message and benefits of the process. It must be more than lip service to appease the field staff of an organization. It goes beyond writing a policy letter and sending it out in hopes the idea will catch on. It requires an executive to invest his or her time in early partnering workshops, train staff, work with industry counterparts and in general set the tone for the implementation of this exciting new program.

In addition to the top managers or executives, it is also necessary to cultivate champions throughout the organization in order for partnering to succeed. First-line supervisors, middle managers and others will play a key role in carrying the executive's message throughout the organization. These champions must be "obsessed" with the process and be personally committed to seeing the principles of partnering implemented properly on every project.

The role of top management in the partnering workshop is discussed further in Chapter 3. It provides the essence of the role of the CEO and other readers in an organization that is implementing this bold and innovative approach to contract administration.

Legal Issues

On one occasion when I was giving a presentation on partnering, one of the attendees stood up and indicated that partnering could not be implemented in his county because it was illegal. He was sure that there was something in statute or code that would prohibit such a relationship. I responded by asking a series of questions. Was there a statute that prohibited working together? Was there one that would stand in the way of cooperation? Perhaps there was a statute that specifically prohibited synergy? Was it against the law to compromise? The answer to each question that day was "no" and has always been "no" each time the issue has arisen.

Part of the beauty of the partnering process is the fact that it can be done wholly within the confines of an organization's procurement code and contract specifications. Nothing you read in this book or hear any knowledgeable partnering advocate say will ever encourage anyone to break the law or do something dishonest. Everything that is done in the partnering relationship must be accomplished in such a way that there is no impropriety or even the perception of wrongdoing.

During the implementation of partnering, there will be those who make allegations of some type of illegal activity. "Why else would the owner and contractor be getting along so well?" they will ask. "If they are not fighting, they must be engaging in some under-the-table wheeling and dealing," they will claim. Hence, always keep the activities associated with partnering process open and fully capable of standing the light of day if necessary.

There are no specific legal requirements that must be dealt with in the implementation of partnering. The need or lack of need for a specification is dealt with later in Chapter 2. The only other legal issue that may need to be addressed is the process for procuring the facilitators and facilities for the actual team workshops. If your procurement process is cumbersome, then it may be necessary to check with the appropriate individuals in your organization in order to avoid a delay of the workshop for a specific project.

The principles of partnering serve as the foundation for a relationship that will ultimately spell the success of the whole partnering team. Historically there may have been those who, over the years, practiced these principles and found them to contribute to their success. Partnering provides the framework under which these principles can be adopted and applied as they have not been before and on a scale never before experienced. Ultimately they will forever change the course of the construction industry.

Time after time the author has heard partnering participants express that they could never return to the "old style" of doing business again. Having experienced the synergy and the power of the partnering relationship, it would be nearly impossible to return to the claims, contention, and problems that were characteristic of the industry in years past.

Chapter 2

Foundations for Success

As an organization begins the implementation of partnering, it is necessary to lay the proper foundation for success. Just as a home or building must be built on a solid foundation, so partnering also requires that certain organizational, specification and educational elements be considered. Properly addressing these areas will provide partnering with the solid foundation it needs to succeed.

Organizational Issues

Partnering champions must ensure that the organization is prepared to function effectively in the partnering environment. A checklist of some of the areas that must be examined is provided below:

- Is top management committed to the partnering process and the cultural changes that must be made?
- Are top managers (to include the CEO) willing to leverage their time in partnering workshops in lieu of the time they would have spent in low-value claim-related activities?
- Is there a procurement process in place that will allow for the hiring of facilitators and the rental of meeting space for workshops and other partnering meetings?
- Does the organization have a list of at least several facilitators who would be available to facilitate the initial partnering workshops?
- Has management determined a strategy for how many projects will be partnered over a certain period of time?
- Has management considered what alternative strategy will be utilized when the industry asks to accelerate the above schedule?
- What record keeping will the organization require to track costs and savings associated with partnering?
- Is the relationship between the owner and industry sufficiently strong to move forward together with the implementation of partnering?
- Will it be necessary to "sell" the partnering concept to some board or commission that has oversight over the owner's operations?
- Has the organization identified one or more champions to carry forward the day-to-day implementation activities of partnering?

- Have employees been sufficiently empowered? (See Chapter 5)
- What contract administration adjustments need to be made to facilitate the implementation of partnering?
- Will it be necessary to make specification changes to your contracts?
- Is staff sufficiently trained in the concept and process of partnering?
- Is industry sufficiently trained in the concept and process of partnering?

These are just a few of the organizational and cultural issues that must be addressed in preparation for the first partnered project. The answers and actions associated with each question will be unique to each organization, but are significant contributors to the success of partnering. This list is not intended to be all inclusive, but rather is intended to provoke thought and to assist in the consideration of the issues that must ultimately be addressed by any organization.

Contract Specification Issues

One of the early issues that arises in the implementation of partnering is whether or not a change in contract specifications is needed. This will largely depend on the organization and its approach to partnering.

A few thoughts on this subject may assist the reader in the resolution of this matter. Partnering must be voluntary in order for it to be effective. The author shudders each time he hears of someone saying that they are going to mandate or legislate partnering in their state. This thought violates the basic premise of partnering as a willing, proactive effort on the part of two or more parties to work in a synergistic environment. While their motives may be sincere, these well-meaning individuals do not yet fully understand the principles of partnering. No amount of legislation or contract language can force synergy to occur.

Given that partnering is voluntary, an organization must decide whether additional contract language is really necessary. The question arises: "What part of this voluntary process is going to require contract enforcement?" Is there a performance measure issue here? (You really cannot force someone to do something that is mentioned in the contract as optional.) Are there quality issues to be addressed? Are we going to punish those individuals whom we perceive to be "bad" partners? A thorough understanding of the process leads one to realize that there is really nothing about the process that requires contract language to enforce. In addition there is seldom anything in an existing contract that would preclude partnering from occurring today. As an organization examines its contracts, it probably will not find any prohibitions to cooperation, compromise, working together or synergy. Without the prohibition of its various principles, partnering can occur freely and naturally in any organization.

When the issue of specification arose, ADOT discussed it with the agency's legal counsel. There was some concern on counsel's part that contractors would not understand the expectations inherent in the partnering process, that somehow a contractor would be surprised by the nature of the partnering relationship. Consequently at counsel's request, a short specification was developed and inserted in each partnering contract (Appendix A). A close inspection of the language reveals that it is simply informational in nature. It should be noted that there are no enforcement provisions typically found in most contract documents. (For comparison, the specification used by the Texas Department of Transportation is Appendix B.)

Going a step further, ADOT determined that a preamble of "good faith and fair dealing" ought to be added to its contracts. These covenants have always existed as basic tenets of contract law, but have been routinely overlooked for many years. ADOT's preamble (Appendix C) reflects this commitment to the parties in its contracts.

In the final analysis, a specification is not needed to establish the kind of relationship that allows partnering to flourish. Successful partnering is founded on personal relationships and the trust and synergy that produce high performing teams. Each organization will have to make its own decision regarding the specification issue. Technically it is not necessary to the implementation of partnering. However a compromise, such as is found in Appendix A, provides contractors some limited information and may serve to appease in-house counsel.

Education and Training

One of the significant lessons ADOT learned in the first couple of years of partnering is the need to invest in education and training. The principles of partnering are so logical and people are so tired of the "old style" of business, that on the surface they readily accept partnering. However the severity and depth of the wounds inflicted over the years are easily underestimated. In spite of an outward acceptance of partnering, it was found that the underlying issues and problems of years past did not go away so easily. A well-thought-out strategy for education and training is essential in dealing with this situation.

Many departments of transportation and other organizations around the country have started their partnering experiences by jointly sponsoring an educational event with owners and industry alike. In the case of Arizona, it was a day filled with information presented by individuals from all segments of the industry who had had partnering experience. This program attracted nearly

800 people over a two-day period with attendees representing numerous organizations both public and private.

From this initial exposure, it was then necessary to communicate the concept of partnering to ADOT's field staff throughout the state. To accomplish this several members of ADOT's top management took to the road and gave similar presentations to all employees involved in the construction process. This was no small effort but has since paid off handsomely in improving the construction practice in Arizona. Contractors have also, to some degree, done the same thing with the same goal in mind.

In addition to formal training, ADOT executives used the early partnering workshops as training opportunities for agency and contractor employees alike. As issues or process concerns arose in these workshop, executives were able to seize these teaching moments and further the education of those present. This commitment to education and training was also instrumental in overcoming some of the uniformity issues that often plague the implementation of such programs. For example at ADOT there are almost 30 Resident Engineers scattered all over the state. It is often difficult to ensure that each is administering the ADOT construction program uniformly on all projects. Uniformity allows contractors to understand the state's expectations and bid projects accordingly.

The initial training received on partnering was a great success, but it is now clear that further training is needed in the field. Much of the partnering process falls to the field staff to implement. After all these are the individuals who have the daily interaction with other team members. Consequently ADOT is now endeavoring to expand the participation of field technicians in the partnering workshops in order to expand their understanding of the process and their "buy-in" to the team and its charter.

Properly thought out, the organizational, specification and education and training issues can be dealt with in such a way as to enhance the early successes of the partnering process. While they may take considerable effort up front, these efforts must be seen as an investment in the future benefits of a new way of doing business.

Chapter 3

The Partnering Workshop

The partnering workshop is an important element of the overall partnering process. A properly planned and executed workshop serves as a foundation for the success of the project team. Conversely, a workshop that fails to establish a cohesive team will ultimately result in disappointment with the partnering effort. Consequently, careful and thorough planning of the partnering workshop must be undertaken in order to ensure this success.

Who Should Attend

The partnering workshop presents a unique opportunity for the project team to meet and prepare for the work to come. One important product of the workshop is a cohesive team. This team should be represented by all parties to the contract who will focus on successful project completion. It is a one-time opportunity for team members to resolve project-related issues without the pressures normally associated with an on-going construction project.

Those planning the workshop should carefully review each party's role in the contract and decide who will be the key individuals on the project. A number of factors should be considered in selecting workshop participants. Some of these are:

- Can the organization being considered impact project completion?
- Who from this organization has decision-making authority for the contract?
- Who from the represented organizations will determine the day-to-day progress of the project?

Individuals or organizations who may impact the success of the project ought to be considered viable candidates for attendance at the partnering workshop. A typical list will include a number of representatives from the owner's organization, principal leaders from the prime contractor's staff, the design project manager and key design team members, individuals from key subcontractors, suppliers and other significant stakeholders. If someone's attendance seems questionable, it is better to have too many people present than to risk excluding a key partner who potentially could be a major contributor to the process.

During the initial implementation of partnering, a CEO/executive-level individual should attend each workshop. While the time investment on the part of these individuals will appear significant, it nevertheless is an essential element of successful partnering.

The CEO's presence at the workshop will accomplish several things. First, it demonstrates corporate commitment to the process. Before they see the vision of partnering, some managers and employees will indicate that they do not have the time to spend in a workshop. Still others will say that pressing matters prevent them from spending so much time on such an activity. However with the CEO present the message is clear—if the CEO is willing to invest the time then they are hard pressed to make the assertion that they are busier than the CEO.

A second reason for CEO/executive-level involvement is related to the establishment of the corporate partnering philosophy. In the initial stages of partnering implementation, many issues arise as part of the new organizational "culture." These issues range from the application of empowerment to new definitions for value engineering to streamlining decision-making processes. They often represent major philosophical changes for members of the partnering team. As an attendee at the workshop, the CEO has an opportunity to provide leadership, direction, clarification and guidance as team members endeavor to establish their partnering framework. When questions arise, the CEO is able to quickly share the organizational philosophy with all the team members so there is no question about the issue and how it will be dealt with.

Another benefit from attendance by someone at the executive level is derived from the association that occurs with his or her counterparts in the industry. The relationships and understandings developed in the partnering workshop will ultimately assist the executive in resolving issues in an equitable manner during the course of the project.

In addition if the organization is implementing partnering among many hundreds of employees, the executive role becomes even more critical. Experience has shown that uniformity of policy is a problem in most large organizations. This being the case, uniform application of the principles of partnering requires central direction in order to align all employees under the same basic philosophical/policy umbrella. Executive involvement allows dissemination of this philosophy to the various branches of an organization as the partnering program begins.

Location and Facilities

The selection of the location for the workshop should be made soon after the contract is signed. A location at or near the actual site of the project is preferred for a number of reasons. First, it is a means of communicating to the field staff that this is not yet another home-office or headquarters program. Second, it also permits team members to go to the project site and quickly check on issues that may develop from the discussion during the partnering workshop. A third reason is that at times it may be advisable to invite members of a local governmental agency or the business community to all or a portion of the workshop. A location close to the project site will maximize the opportunity for these individuals to attend.

With the workshop location selected, the choice of a facility is next on the agenda. There are two options from which to choose. First, it could be held in the office of the owner or perhaps that of the prime contractor. Second, a neutral facility could be selected. The office of the owner or prime contractor has the advantage of being readily available and it does not cost the partnering team anything. The temptation for some might be to save the extra money and hold the workshop in the already available room at the owner's or contractor's office. This should be avoided. The neutral facility results in far better results for the overall partnering process.

A neutral facility provides an environment where team members can come together in non-threatening surroundings and develop into the project team. Previous barriers to teamwork such as project pictures or other typical office decor are not present, so the formation of the team can proceed uninhibited.

Another advantage of a neutral facility is that it removes the team members from their normal work environs and the inevitable distractions and interruptions that come with them. Team members should be encouraged to delegate all non-partnering duties to others during the workshop so they may fully concentrate on the partnering process.

A typical facility for a partnering workshop consists of a large meeting room that provides adequate working space for all workshop participants. This room will serve as the center of activity for all of the elements of the workshop. In addition to the large room, two to five smaller breakout rooms are necessary for small-group work. As the workshop progresses, some activities occur in the large room with all team members present while others take place in smaller groups in the other rooms.

In many cases the selected facility will be a local motel or hotel near the project site. This provides out of town participants a convenient place to stay close to the workshop meeting. Sometimes a restaurant will have room to accommodate a partnering group. In small towns, a local club or even a church may serve as host. An organization may have to be very creative to maintain its resolve to hold workshops close to the project site. The effort expended in this regard is well worth it.

The selection of a motel or hotel offers additional advantages over those already enumerated. Foremost is the fact that they will generally have dining facilities large enough to provide lunch and perhaps dinner to the team members during the course of the workshop. While this may seem insignificant, the value of "breaking bread" together as a team must not be underestimated. Again, organizations must avoid the temptation of dodging the nominal expenses associated with the luncheon because it is an integral part of the team-building process.

Facilitation

The partnering workshop is traditionally a very intense period of one to several days. Many times it is also the first time the parties have met together in a non-confrontational setting and worked together on a problem. Issues will arise and solutions will be sought. The old style of solving problems must be abandoned in favor of the partnering approach. These habits are not easily given up. Given these circumstances, the use of outside facilitation is absolutely essential.

A facilitator assists the team in putting together its goals and objectives, drafting the charter, learning to work together as a team and resolving the many project related issues that will come up as a part of the workshop. The facilitator will keep the meeting on schedule and will adjust various elements so as to develop the strongest partnering team possible. He or she will be invaluable in making the most productive use of the team's time through the course of the workshop.

Outside facilitation is a must in the initial stages of partnering implementation. This is not to say that some organizations do not have qualified internal personnel who possess the necessary skills. In the future these individuals might serve as facilitators after the partnering process has matured both within the organization and within the industry. However their effectiveness is limited while team members are still getting used to the idea of working together. In the initial stages of partnering when the foundations of trust have yet to he developed, these internal facilitators will be seen simply as a representative of the other party.

There are no established qualifications for partnering facilitators. After over 100 partnering workshops and numerous facilitators in Arizona, there appears to be no one set of qualifications that guarantees an effective facilitator. Rather, an organization must review the credentials of each individual and make a judgment as to potential effectiveness. Typically individuals with experience in team building, productivity and conflict resolution may make good facilitators. Some facilitators have an industrial engineering background and others have experience in legal matters, while still others have been working with the construction industry on other team-building issues. A good source for information on available facilitators would be other organizations involved in partnering in the same locale or region.

The facilitator should assist the key managers from each organization in developing the agenda for the workshop. He or she should lead the workshop through some portions of the agenda and allow the team to chart its own course during others. Another function of the facilitator is to produce the written record of the workshop and furnish it to the team participants. This permanent record can then serve as the basis for future reference should issues arise during the course of the project.

Agendas that Work

In order to maximize the value of the partnering workshop, several key elements are essential. The effectiveness of the process will depend on how the workshop is structured around these elements.

Exhibit 1 is a typical agenda for a two day workshop. The times are approximate for each activity and will vary depending on the project and the facilitator. In addition, as the partnering process matures in an organization, there will be opportunities to adjust the agenda to reflect this fact.

Exhibit 1 Partnering Workshop Two Day Agenda

Day 1

    8:00 Welcome and Introduction of Participants
    8:30 Overview of the Workshop
    8:45 Team Building Exercise
  10:00 Organizational Goals and Objectives
  11:00 Common Goals and Objectives
  12:00 Luncheon
    1:00 Charter Development
    2:00 Project Issue Identification and Resolution
    5:00 Adjourn
    6:30 Dinner (Optional)

Day 2

    8:00 Review of Day 1; Overview of Day 2
    8:30 Issue Resolution Continued
  11:00 Team Presentations of Issues
  12:00 Luncheon
    1:00 Issue Escalation
    2:00 Joint Evaluation
    2:30 Charter Signing
    2:45 Wrap-Up
    3:00 Adjourn

On smaller or less complicated projects it is possible to adjust the agenda to reflect a shorter workshop. Exhibit 2 is an example of how this might be done. This approach is also effective on a larger project where the team has worked together before and less time is needed for the team building part of the program. At all times, it should be understood that each element can be adjusted to suit the specific needs of the project in order to maximize the effectiveness of the process.

Exhibit 2 Partnering Workshop One Day Agenda

    8:00 Introductions and Welcome
    8:15 Team Building Exercise
    9:00 Organizational Goals and Objectives
    9:30 Common Goals and Objectives
  10:00 Charter Development
  11:00 Issue Identification and Resolution
  12:00 Luncheon
    1:00 Issue Resolution Continued
    2:00 Team Presentation of Issues
    3:00 Issue Escalation
    3:30 Joint Evaluation
    4:00 Charter Signing
    4:15 Wrap-Up
    4:30 Adjourn

Team Building

Most of the elements of the partnering workshop are covered in some detail in other chapters, and they will not be addressed here. However team building is one area that deserves more attention at this time.

Team building is generally the first element of the partnering workshop. It is intended to "break the ice" for the workshop participants and will set the stage for the other team activities through the remainder of the meeting. These team building exercises will take the form of games or assignments that are designed to cause workshop attendees to work together to solve a given problem. At one workshop puzzles were used during this part of the agenda. Groups were given puzzles to put together, but were not permitted to talk to anyone else in the room. It was a timed event with a reward going to the winner. Unknown to the participants, there was one piece from each puzzle that had been switched with that of another group. It took a while until all of the groups realized that by working together, instead of separately, all of the puzzles could be completed at the same time and everyone receive the reward. Other workshops have included situations or scenarios where groups have to work together to extricate themselves from some difficult situation. These exercises are fun and should be viewed as a valuable part of the partnering workshop.

Each facilitator will have his or her own style and approach during the team-building portion of the workshop. This variety is good for two reasons. First, it allows each facilitator to use the skills they find most effective in this part of the workshop. Second, if team members attend these workshops in a short period of time, doing the same team building exercises would be redundant.

Some people within the various organizations will feel the team-building activity is a waste of time and that the exercises or games are silly or foolish. They would prefer to skip this part and move on to more "important" elements. Team members must see past the games or exercises themselves and catch the vision of the transformation that is taking place in the various team members. The product of the team building portion of the workshop is tangible, and it is essential to the success of the rest of the workshop.

Issue Identification and Resolution

A very productive part of the workshop will be the time spent identifying and resolving issues that will impact the project. Regardless of the most diligent efforts to produce quality plans and specifications, there will always be questions or concerns that must be addressed to ensure a successful project. During the partnering workshop, there is ample time to bring these issues forward for the partnering team to resolve.

Typically all of the team members have issues to bring up at the workshop. The contractor needs details on a part of the plans. The owner needs information about how the contractor will approach a certain part of the project. After all of these issues are listed, the team will generally break down into smaller teams that can tackle each problem and come up with solutions or at least a plan of action for resolution. Some issues will not be resolved at the workshop for any number of reasons, but a plan with a time-frame and responsible party are the minimum required on every issue. This process of resolving problems causes a solidification of the team that is tangible and will serve as the basis for future problem resolution.

The partnering workshop is an excellent tool for the implementation of the partnering process on a construction project. Team members will come away elated at the prospect of doing business in such a cooperative manner.

However, problems will soon arise on the project as they always have. The difference now is that the team is prepared to deal with those problems as they never have been before. This is the ingredient of partnering that makes it the great tool that it is.

Chapter 4

Charter Development

One of the most important products that results from a partnering workshop is the team charter. It represents the common commitment of all team members and serves as a philosophical guide to the team throughout the project. If properly prepared and committed to by the participants, the charter becomes a powerful tool in resolving problems and issues as they arise during the course of the project.

Goals and Objectives

The organizational goals and objectives portion of the workshop, coupled with the time dedicated to developing the team's common goals are preliminary steps to drafting the team charter. Broken down into two elements, this part of the workshop allows individuals and organizations to reflect on what is truly significant to them from both project-specific and global perspectives.

This portion of the workshop is needed so that each participating organization may review its specific goals and objectives for the project. Contractors find that their goals go far beyond "earning a profit." Owners find that they are interested in more than just getting a cheap project completed early. This session provides time for introspection on the issues of quality, safety, training and education, organizational culture and procedures and strategic goals for all team members.

Once the respective organizational goals and objectives are identified, the focus of the workshop then moves to the consolidation of a list of those that the team will share for this particular project. This will be accomplished through any number of processes that the group and the facilitator find effective. Often the process is as simple as each organization putting their respective lists on the wall, side by side. Then team members match those that are common. This consolidated list becomes the project team's common goals and objectives. Team members who are new to the partnering process will marvel at how parties with such diverse stakes in a contract can ultimately share so many of the same concerns. This realization has a powerful psychological impact on the team members and further assists in the team building process.

The listing of team goals and objectives will generally include issues of quality, project schedule, project administration, value engineering, safety and claims. Each project team will have a set of specific issues unique to the current project. In one early workshop the following goals were developed:

Contractor	ADOT
Finish in 6 months or less	Timely completion
Successful partnering	Successful partnering
Generate estimated profit	Budget—save 5%
Quality work	Quality project
Safe job	Accident-free job
Efficient wall construction	Good community relations
Streamline contractor QC	Paperwork reduction

While the words are not necessarily the same, the team came to realize that they did share many common goals and that they could work together towards those goals.

An important element in developing the common goals and objectives is that they must be specific and measurable. Too often it is easy to identify a goal and not take the time to be specific about its performance. For example, a team may set a goal to "have a safe project." This is certainly a goal that everyone in the industry should strive for. However, it is not specific enough to allow team members or management to evaluate the team on achievements related to this goal. A better goal in the area of safety would be "no lost-time accidents." This gives the team members a measurable goal to work toward. Members can clearly compare their performance with this goal.

The Team Charter

With the team's goals identified, it is now time for charter development. The charter represents commitment on the part of all team members to the partnering process and more specifically to the upcoming project. A project charter is composed of several elements. The first is a short mission statement identifying the team, the project by name and the team's superordinate objective for the project. Next is a listing of the team's common goals and objectives as previously developed in the workshop. Finally, many charters are completed with a closing statement indicating the commitment of team members to this partnering effort. The bottom portion of the charter is usually reserved for the team member's signatures. A sample charter is found in Appendix D.

The mission statement is simply a one- or two-sentence declaration of the overall purpose of the project. For example, perhaps the purpose of a project is to build a bridge over a river or to reconstruct a traffic interchange. This should be stated. In addition, the partnering team should be identified. Many charters start out with statements such as "We, the partners on the Dunlap Traffic Interchange project. . . ." The mission statement also includes a reflection of the team's superordinate objective for the project. What is it that the team is ultimately going to accomplish? It might be that the superordinate objective is to provide an all-weather crossing of the Gila River on State Route 587. This helps the team focus on the real intent of the project.

After preparing the mission statement, the team should insert the goals and objectives developed earlier in the workshop. In addition, the members may want to conclude with another statement of commitment to the partnering team.

In Arizona it was decided that charters should be personalized to the specific project team, so company logos are often added to the border of the charter. This provides a sense of ownership to the charter and enhances the commitment felt by the individual team members. The signing of the charter can be a momentous culmination of the partnering workshop. If the CEOs from several of the organizations are present, they may choose to make the signing a symbolic activity that demonstrates their dedication to the principles of partnering on this and other projects.

After partnering over 100 construction projects, the author can make some interesting observations concerning the partnering team charter. First, individuals may hesitate to sign the charter at the beginning of the partnering process. It makes people nervous to sign documents that appear so official. This is particularly true of organizations that have done little about empowerment. Sometimes individuals who participate as representatives of organizations that are not formal parties to the contract are concerned about signing the charter. The charter is not a legally binding document and does not circumvent or otherwise take the place of the contract documents.

The truth is all workshop participants should sign the charter regardless of whom they represent. It does not bind their organization legally any more than if they do not sign it. Signing the charter, however, demonstrates commitment that should outweigh such concerns. More important is the demonstration of commitment that the signing of the charter signifies. For those who may yet be unsure about signing the charter, the following is a partial list of those who have already done so in Arizona:

Designers and architects
Subcontractors
Suppliers
Federal Highway Administration (FHWA)

Bureau of Reclamation
U.S. Forest Service
Utility companies
Local governmental agencies
Tribal governments

There should be no concern about signing the charter after one understands the purpose and role of this document in the partnering process.

The partnering charter becomes the most important piece of paper on the project. It becomes the guiding influence when other forms of communication break down. The personal commitment they make by signing the charter motivates them to a higher level of contract compliance. It may not be a legally binding contract document, but the charter has a moral influence that is unmistakable in its impact.

The time spent in the development of the team's goals and objectives, coupled with the writing of the charter are important elements of the workshop process. They should not be rushed or taken lightly. In addition the temptation to copy a previous charter should be avoided at all costs since the team would have no real ownership of the duplicate charter. When properly completed, the charter will be a valuable tool as team members strive to work together in the accomplishment of significant efforts during the project.

Chapter 5

Empowerment and Issue Resolution

One of the key elements of the partnering concept is empowerment. It describes the delegation of authority and a shift in traditional corporate philosophy that may be new to many organizations.

Concept of Empowerment

Empowerment is one of the words that has become part of our 1990s vernacular. It is used frequently by firms that are heavily involved with Total Quality Management. As with any popular concept, there are times when the term *empowerment* is improperly applied. True empowerment is a powerful tool in the partnering tool chest. It has the potential to make a significant difference in the success of an organization's partnering effort.

Empowerment is the delegation of authority and responsibility to the lowest possible level in an organization. It is not an abrogation of responsibility, but a transfer of authority and accountability to individuals throughout an organization to facilitate timely decision making. It is based on the premise that the individuals closest to the problem are best equipped to make related decisions.

An organization planning to empower its employees must examine what authority it can legally delegate. Secondly, it must decide what is the best level at which to delegate that authority. For example, in every state there is a statute that details which engineering drawings must be signed and sealed by a professional registrant. In this case it would be improper and illegal to attempt to empower a draftsman to sign and seal drawings. It would be appropriate to have a design firm principal empower a professional registrant to seal a set of plans for which he had been responsible.

Authority cannot be delegated without accountability. An individual must be accountable for the decisions he or she makes. Otherwise, employees could make irrational decisions knowing they would not ultimately be held accountable for them.

Empowerment must be formally established in an organization. All levels should understand the various degrees of authority and their respective boundaries. Lack of clarity in this area results in confusion and misunderstandings as individuals perform the duties of their respective positions.

The foundation of empowerment is the assumption that the organization is made up of competent, reasonable and rational individuals who are going to make good, sound decisions on nearly every occasion. True there may be times when they will make a decision a senior manager does not fully agree with. On these occasions the natural temptation will be for that manager to change or override the subordinate's decision. This reaction should be avoided at all costs unless someone's life or property is in jeopardy. A questionable decision by a subordinate presents the manager with an opportunity to train the subordinate so he fully understands what the course of action should have been. However, great care should be taken to prevent the subordinate from feeling unsure about making future decisions. Nothing will discourage empowered employees more than being "slam dunked" by a manager after a particular decision.

The Arizona Department of Transportation made empowerment an important part of its partnering effort from the very beginning. Prior to partnering agency policies established authority levels for construction contract modifications at $15,000 for Resident Engineers and $50,000 for District Engineers. All other changes required approval by the State Construction Engineer. The system was cumbersome and did not reflect ADOT's confidence in its field engineering staff.

A study done of the many changes made on ADOT's construction contracts revealed that about 85% of the total had a value less that $5000. In addition almost all of the remaining changes had a value of less than $200,000. The study also revealed that by the time the State Construction Engineer finally signed or approved the change document, the work had been completed for several months. Ultimately it was a process that needed to have the "value-added" test applied to each step.

In the spirit of empowerment, ADOT changed its delegation of authority for contract modifications to $50,000 and $200,000 for the Resident Engineer and District Engineer, respectively. Now only those few documents that exceed the $200,000 threshold must come to the headquarters for approval. In addition streamlined procedures have been implemented to deal with the many low-dollar changes that continue to occur.

These changes have been in effect for about 2-1/2 years. During that period of time, ADOT's field engineering staff has seized this opportunity to make timely decisions for the benefit of all the partnering team members. ADOT management is very pleased and proud of the way this new authority has been accepted and applied.

Issue Escalation Ladder

The concept of timely decision making is related to empowerment. One of the products of the partnering workshop is the issue escalation ladder which facilitates the process of making timely decisions. This is the name given to the plan for problem resolution on a partnered project. It is a unique document that will facilitate the quick resolution of many issues arising during the course of a project.

A typical issue escalation ladder is found at Exhibit 3. It is created by the partnering team during the workshop before the beginning of the project. The participants start by listing the chains of command for the owner and prime contractor. Next, they decide which individuals in each organization will deal with one another during the course of the project. In Exhibit 3 Smith and Davis are on the first rung of the decision-making ladder. They report to Johnson and Wood respectively, who then report to their managers and so on. Ultimately, the top level of the issue escalation ladder consists of the CEOs of both the owner and prime contractor. The spirit of the process dictates that no decision or dispute should go past these two individuals for resolution.

In many issue escalation ladders, the time frames are delineated in order to facilitate the decision-making process. Exhibit 3 shows how this is often done. The significance of the time frames is that each layer in the ladder has a specific time in which to act before the issue must be automatically escalated to the next level. This prevents an issue from simmering at one level while progress on the project is impacted.

Issues can be escalated for a number of reasons. One is that a person at a particular level may not have the authority to deal with the problem at hand. Perhaps the concrete curing process does not meet the specifications of the contract. The inspector either does not have the authority to modify this requirement or feels uncomfortable making such a decision. Thus it must be escalated to a higher level in order to have it resolved.

Another reason is when the two individuals at a particular level cannot agree

Exhibit 3 Issue Escalation Ladder
Sample Project: Casa Grande Traffic Interchange

Owner	Time	Contractor	Level
Banks	1 Day	Hunter	CEO/Executive
McDonald	1 Day	Fischer	Vice President
Hughes	4 Hours	Leavitt	Project Manager
Johnson	4 Hours	Wood	Project Superintendent
Smith	2 Hours	Davis	Foreman

on a resolution to a problem. Their respective positions may be so far apart that compromise would be inappropriate for either one. Consequently the issue is escalated until individuals at a higher level in both organizations are able to resolve the matter. Sometimes it is necessary to raise the issue to a level where a manager can see the "big picture" for his or her organization.

Several guidelines must be followed in order for the escalation ladder to work. First, issues must be escalated by both organizations at the same time. No one is allowed to go behind another's back. Second, no one is allowed to sit on a problem and prevent a solution. Either way, they must deal with it and resolve the issue or they are obligated to escalate it to the next level. "No decision" is an unacceptable course of action under the partnering scenario.

There are a number of barriers to the success of the issue escalation ladder. Primarily, we must change the paradigm about admitting to our supervisor or manager that we cannot solve a problem. The author's experience has shown that employees are at first reluctant to escalate problems because of this paradigm. Nevertheless with the right encouragement and coaching they should begin to realize that escalation is often the most responsive course of action they might take. The system begins to work when team members come to realize that problems must be dealt with rapidly in order to maintain commitment to the partnering team.

Detractors to the escalation process often express concern that the ladder will only promote the resolution of problems at the highest level in an organization. They assert that the CEO/Executive whose name appears at the top of the ladder will be involved in most of the decision-making that occurs on the project. This has not occurred. The author's experience has shown that empowered employees rise to the challenge and make sound decisions. On many partnered projects the teams address the issue of escalation in their charter. It is not unusual for a team charter to include a goal statement such as "Resolve 95% of all problems at the project level" or something similar. When given the authority, employees are more than willing to make decisions with which they feel comfortable. The fact is, after two years and over 100 partnered projects, only about a dozen issues have been escalated to the top of the ladder in ADOT. By and large project team members are resolving problems and meeting their commitment to fellow team members.

Some detractors state that empowerment of the owner's employees means that they will now never say "no," that the owner will always have to give-in due to the commitment in the charter. This is not true. The issue escalation ladder provides employees with the opportunity to make the best decision. However, if they are uncomfortable with a particular issue, they can escalate it and allow someone with higher authority or greater experience to resolve the problem. "No" may still be a correct response to a request by the other party. It may result in the issue being escalated for the next level to deal with. The author's experience has shown that the issue escalation ladder tends to reduce or eliminate the petty, ridiculous or unreasonable requests or responses since the possibility always exists that the issue may be escalated.

An interesting example concerning empowerment occurred on two projects that were being constructed simultaneously in the same town in Arizona. The prime contractor and the state crews were basically the same on both projects. However the key difference was that on one project the contractor had an experienced and empowered superintendent. The other project was led by a less-experienced superintendent with little authority. The results were not surprising. The second project struggled because the less-experienced individual was unable to make timely decisions and this adversely affected the outcome of the project. The partnering process was not a failure in this case, but the results were not what they could have been. Empowerment must occur in both the contractor's and owner's organizations to be truly effective in the partnering environment.

Empowerment and the issue escalation ladder are essential ingredients to the success of the overall partnering process. All organizations involved in the project team must be committed to empowerment or the process will fall far short of its full potential. Proper use of the issue escalation ladder will facilitate timely decision-making on the project and provide an environment where all of the team members can succeed.

Chapter 6

Expanding the Partnering Team

A long-held paradigm on construction projects has been that the owner and the prime contractor were the only important parties to the contract. So entrenched was this paradigm that in many cases owners would choose to have no communication or involvement with subcontractors and suppliers due to the fact that they had no direct contractual relationship. In addition this same attitude caused owners to prevent contractors from communicating with designers for year that contact might prejudice their contract administration activities.

This lack of communication and failure to recognize the role of all project participants contributed to the negative trends characteristic of the construction industry for so many years. The partnering concept, however, challenges this paradigm by elevating subcontractors, suppliers and designers to full-partner status on the team. This results in stronger, more effective teams and more successful projects.

Subcontractors

Prior to partnering, the subcontractor was given second-class citizenship at best. Many times subcontractors had no say in major project decisions that greatly affected their operation and ultimately their bottom line. Problems with prompt payment, clear direction and fair representation were common and reinforced the traditional position of the subcontractor on the project.

With partnering the subcontractor's role has changed. This change is obvious from the outset when the subcontractor is invited to participate in the partnering workshop and signs the charter along with all other team members. The views and concerns of these subcontractors are heard and addressed because the other team members recognize that a subcontractor's performance can impact overall project completion. The issue of prompt payment is less frequently a problem because subcontractors, contractors and owners are working more openly and honorably with one another and want their counterparts to succeed.

However when subcontractors are given greater respect and participation, they also incur greater expectations and responsibilities. Subcontractors must become absolutely reliable in keeping their scheduled commitments. Knowing that other team members are depending on them to perform quality work in a timely manner should cause subcontractors to be even more effective than ever before.

Subcontractors should feel more comfortable suggesting value engineering proposals to the team in the partnering relationship. They know that no legitimate idea will be prematurely discarded by the team. In addition they realize that savings generated from a value engineering idea will be appropriately shared among the team members.

Subcontractors should revel in this new-found status and seek to make their role important to the success of the team. Ultimately they will be more effective and profitable and will gain a long-desired respect in the industry.

Suppliers

Much of what has been said about subcontractors applies to suppliers as well. The main difference is that suppliers were relegated to an even lower status than subcontractors in the pre-partnering relationship.

There are generally a host of suppliers who will provide products and services for various elements of a project. As the partnering team is being assembled, key suppliers should be identified as team members. They may be selected based on the size, dollar value, volume, or the critical nature of their product. Regardless of the criteria, significant suppliers must be integrated into the partnering team as full-fledged participants.

Much like subcontractors, suppliers will enjoy both the greater benefits and higher expectations inherent with their new role. The benefits include a greater opportunity to control their destiny within the project team. The expectations involve having an absolutely reliable product delivered on time for installation by another team member.

Designer

The new-found role of designers is also significantly different. Prior to partnering, designers typically turned a set of plans over to the owner and then had only limited involvement during the construction phase of a project. Seldom did they have an active role in the field activities of a project they designed. In some cases it was a matter of money, the owner wanted to save on the final cost of a project. In other cases, the owner was afraid that a close relationship would allow designers the opportunity to cover up some design error.

In the partnering process, designers will quickly sense the magnitude of their new role. The designer will be invited to the partnering workshop and will find himself or a herself a central figure in the issue resolution portion of the session. For this reason, the individual who attends the workshop should be empowered to make decisions that will impact the design of the project. Typically attendance from the design/architectural firm includes the project manager and key contributors to the project design. If value engineering proposals are brought forward by team members at the workshop, they can be quickly referred to the designer for review and recommendation. Often the designer will leave a workshop with a list of issues for analysis and response based on the request of the other team members.

In addition to playing a significant role in the partnering workshop, the designer should also be very active in the field work on a project. This is not to say, that the designer replaces the owner's resident engineer or project manager. Instead the designer becomes a Minute-Man resource to the field staff. To those who are new to this concept, the author suggests you consider the designer as an extension of the resources the resident Engineer has at his or her disposal.

On most projects the designer should visit the site frequently and be prepared to provide guidance or answer questions posed by other team members. The designer's routine visits will facilitate timely decision-making and assist all team members to be more effective in their respective roles.

The partnering relationship with the designer requires some adjustment on the part of both owner and designer. The nature of the work requires the designer's key managers and sometimes a firm principle to be available to make the kind of timely decisions that will come about on a successful partnering project. This may require some organizational or procedural changes on the part of the owner. It may be necessary to change the payment structure for designers in order compensate them for this increased level of activity. In ADOT's case, the normal pre-partnering compensation package consisted of a set rate per hour of time spent by the designer regardless of who was involved from the designer's firm. When ADOT started partnering, it found that the state was demanding more time from the project manager and firm principal than was generally the case in the past. The fixed hourly rate often did not cover the out-of-pocket costs incurred by the designer in performing this new role under partnering. Consequently in the spirit of fairness, ADOT changed its compensation process to pay the actual rate per hour for the time of the individuals working on the project as part of the partnering team.

The Complete Team Approach

The partnering team is not unlike a sports team. Just as a baseball team would never take the field without a catcher or center fielder, so the partnering team should never attempt to field a team without key members present. The complete partnering team includes the owner, prime contractor, subcontractors, key suppliers, the designer and any other entity that might determine the success or failure of the project. To do otherwise would be to set a course away from partnering and the opportunities it affords those who embrace this process.

Chapter 7

Partnering during Construction

The newly established partnering team needs nurturing in order to flourish. A common misconception of some users of the partnering concept is that once the workshop is over you have "partnered" and that is the end of the process. Then during the project they start to wonder why partnering is not working as well as they expected. They do not realize that simply having a workshop does not guarantee that they are in fact partnering. A brief review of their process will generally reveal the absence of continuous application of the partnering principles.

Keeping the Partnership Alive

A team can do a number of things to maintain the feeling of teamwork and unity that came out of the workshop. It is a good idea to provide each team member with a copy of the charter. Copies can be reproduced on parchment and framed for display in the offices of the various member organizations. They should be displayed in conference rooms and other routine gathering places as reminders of the commitments made before the project started. Some teams even choose to review their charter at each weekly progress meeting in order to set the tone for the discussions that will follow.

Another tool that teams often use is a picture taken at the close of the partnering workshop. Each person is provided a copy of this picture and it is also typically displayed adjacent to the charter. To the question, "Why a team picture?" the response is "What successful sports team does not have a team picture?" In the partnering team picture, it is hard to identify who is from which organization. Over the course of the project, the blending of the team is quite striking. The transformation of a group of individuals representing a diverse collection of organizations into a team is one of the exciting aspects of the partnering process. Ultimately the picture is another reminder of the workshop wherein the team developed its common goals and objectives for the project.

Occasionally team members may need to gather during the course of the project in a less formal setting such as for lunch or dinner. This allows the team to separate itself from the dust and dirt of the project and truly evaluate the team's effectiveness. A powerful component of such a meeting would be the senior partners or CEOs attendance and re-expression of their commitment to the team and its success.

Addressing Problems as a Team

As partnering teams mature, they move away from the "my problem—your problem" syndrome to an "our problem" approach to issues. Understandably for some team members, this can be a difficult transition. Team members must make the conscious mental shift that as problems or issues arise on the project they should be handled by the team. Regardless of who is at fault, the team faces problems that may impact the success of the project. The best approach to resolution is for the team to work together synergistically. This does not excuse or absolve the party at fault, but does result in the best and most economical solution to the problem. The true team approach will find all of the team members working toward this goal.

An early lesson in the implementation of partnering occurred on a project in northern Arizona. The contractor was constructing a rather long bridge structure on drilled caissons. In one case the caisson was drilled about 18 inches off center. Some discussion among the field staff from both ADOT and the contractor ensued, but no resolution was agreed upon. Prior to the partnering team addressing the problem, a foreman from the contractor chose to try and adjust the rebar cage with a crane. Ultimately the cage was damaged when it was bent over by the crane and the team was now forced to deal with an even more significant problem. A great deal of rework was required to repair the damaged caisson, which was costly in both time and money for the contractor.

Citing this example is not meant to fault the contractor's foreman. This was one of ADOT's first partnered projects. In the case of the caisson, it would have been a matter of a minor change in the rebar detail in the pier cap to solve the problem if the foreman had been willing to come forward and trust the team. The error could have been repaired for a fraction of what it did cost the contractor in both time and money. However it does demonstrate an important lesson for everyone embarking on the road to partnering. Regardless of the amount of "preaching" we do about high trust culture and the other principles of partnering, it does take a certain amount of confidence in your team in order to come forward and put your destiny in their hands in order to resolve a problem. Whether you are the owner who has discovered an error in the plans or the contractor who has made a mistake in your work, it takes courage to make the leap of faith necessary to confess your sins to the other members of your team. No matter who was at fault, it was a costly lesson for one of our partners in the implementation of partnering.

Revitalizing the Partnership

In spite of the best of intentions, there will be projects where relations and team-work are not what they ought to be. The problems may be attributed to many things. Perhaps there are some basic personality conflicts. On the other hand, problems may stem from adding new team members who did not participate in the original workshop. Regardless of the cause, it is the responsibility of the team leaders to move quickly in addressing this serious issue.

This condition has occurred on a handful of partnered projects in Arizona. One technique that was used to set it back on course has become known as the reaffirmation meeting. This meeting is a gathering of all team members at some mutually agreed upon location to discuss their commitment to the charter. Usually the meeting is jointly run by the owner and prime contractor CEOs. The problem is discussed in general terms avoiding accusations and, in many cases, specific issues. More important are the expressions of feelings on trust, cooperation and teamwork. The CEOs guide the discussion such that team members sense the team spirit which existed at the original workshop. If the CEOs feel uncomfortable in doing this, a facilitator may be called in to assist. Generally, a reaffirmation meeting ends with each team member affirming his or her renewed commitment to the charter, to the team and to the success of the project.

From being involved in numerous teams the author has observed a hesitancy to hold the reaffirmation meeting. Teams put it off thinking that their problem is not bad enough or that it will improve. Sometimes they feel that there just is not enough time left in the contract to make a difference. Other times they are hesitant because someone might find out they were not being true to the partnering charter and this will cause embarrassment. Problems just do not get better with time. When in doubt, hold the meeting.

The Follow-Up Workshop

On some projects it may be advisable or necessary to have a follow-up work-shop for two reasons. First, the duration or complexity of the project may be cause enough to gather the team back together. On the other hand, there may be enough changes in the make-up of the team that it makes sense from a team-building perspective.

The follow-up workshop can be tailored by the team to suit its specific needs. If the addition of new team members is the issue, time should be spent in the team-building process again. If there are significant technical issues to be addressed, time should be focused on issue resolution.

A follow-up workshop can last from a few hours to one or more days depending on the reasons for gathering. Again the tendency is for teams to delay holding the follow-up workshop, often saying they can not afford the time away from the project. However, if the team is at risk, they cannot afford not to take the time.

Active and successful partnerships just do not happen, they are created and nurtured. It takes an investment of time and effort to bring about this success. This commitment of time and energy is certainly preferable to time and man-power spent dealing with conflict, claims and litigation. Those responsible need to continually monitor the team's progress and determine what steps are necessary to maintain the momentum of the original workshop. The partnering team that is left alone without care and feeding will wither just as a plant that is denied proper nourishment.

Chapter 8

Value Engineering

Value engineering (VE) is a concept that has been around the industry for many years. Unfortunately for a variety of reasons, it has seen only moderate use on construction projects. In fact not every Department of Transportation even includes it as an option in their highway construction contracts.

Value engineering has historically offered the contractor an opportunity to propose alternate means or materials to achieve a product of equal or greater quality and value on a project. Often, however, owners have not encouraged the submission of VE proposals because they do not want to hassle with them. On the other hand, designers were often offended by the idea that someone would come up with a better approach to a project that they had spent considerable time designing. Consequently, contractors were hesitant to expend time and resources to propose innovations to the project plans. As organizations catch the vision of partnering, the concept of value engineering begins to take on different attributes and has far reaching implications to owner and contractor alike.

Value engineering offers the partnering team a great opportunity to allow this synergism to take hold and produce innovative results. The skills and innovation that contractors, designers and owners bring together set the stage for exciting VE opportunities on any given project.

The partnering workshop provides an excellent opportunity to allow this synergism to take hold and produce innovative results. The skills and innovation that contractors, designers and owners bring together set the stage for exciting VE opportunities on any given project.

The partnering workshop provides an excellent opportunity for initial discussions of VE ideas. Some large projects will even have a separate session just to sort through and prioritize value engineering issues and to develop appropriate action plans for implementation. Where the number of VE ideas is small or the project simple, the VE portion of the workshop can be incorporated into the time dedicated to issue resolution.

Value engineering within partnering is modified to make it consistent with the one-team approach to the project. VE proposals are no longer considered solely the contractor's idea, something to be tolerated: instead they have become team proposals with opportunities for the team to excel. This is how the terminology *value engineering joint proposal* (VEJP) was developed.

A VEJP generally starts out as an idea from one of the project team members. It may come from the contractor, subcontractor, owner or designer. Regardless of the source, the team is committed to dealing with the idea in a timely and responsible manner whether or not it ever becomes a successful VEJP.

Once the idea is submitted, the team assembles to consider its value and potential for savings. The team is committed to making an initial decision very quickly in the process before any team member expends a significant amount of money. The team may also decide to perform more analysis prior to making a decision but should remain focused on the need to make an early decision on a VEJP.

The Value Engineering Paradigm

Some owners are already willing to consider value engineering proposals set forth by contractors. Their contracts often include a mechanism to share some of the savings with the contractor. Usually if a contractor submits an idea and it is implemented, the owner will split the savings based on a predetermined formula such as 50–50 or 60–40. On the other hand, the owner has traditionally reserved to himself all of the savings if a member of his staff generated the idea. This is the value engineering paradigm.

As an organization and its members catch the vision of partnering they begin to see the failure in logic of this paradigm. "If it's my idea (speaking as the owner) then I am happy to keep all of the savings. However, if it is your idea (the contractor) then I will be happy to share the savings with you." This approach does not fit the spirit of partnering and the principles that make partnering successful. Realizing this paradigm exists is the first step toward change. The second step requires a shifting of paradigms and a rising above past feelings.

The new philosophy behind value engineering is that after the contract is awarded, all the savings are shared regardless of whose idea is implemented. This is the total team approach and the product of the power of synergism that partnering cultivates.

Of course, there will be owners who don't want to share the savings generated by their ideas. To this, the author responds with the following comments. First, to espouse such a philosophy is to endeavor to stifle the very synergy upon which the concept of value engineering is based. What motivation is there for a contractor to support or enhance an idea set forth by the owner's representative? If the owner tries to force the issue, the contractor has a multitude of opportunities to ensure that the owner's benefits are either reduced or never realized. It would be better to have a willing partner in the process so that all parties involved in the contract can realize the benefits of value engineering.

This brings up the second point to be considered on this issue. Few ideas, regardless of their source, can be implemented without some measure of cooperation on the part of all parties. True, an owner can "order" a contractor to perform certain items of work. However, the best price and the best product will never be achieved without the synergistic cooperation of all parties to the contract.

Another matter to consider is the fact that the owner has typically been involved in the development of the project for months or years prior to the start of construction. This period of time has afforded the owner every opportunity to provide the most innovative set of plans possible. Many times VE ideas from the owner will change the very nature of the contract upon which the bid was prepared. This being the case, the issue becomes one of fairness and honor in the contractual relationship. To exclude owner generated ideas from the VE process during construction is to compromise the integrity of the partnering relationship.

Value engineering and the synergy that creates it are wonderfully powerful outcomes of a truly successful partnering team. They provide an even greater foundation for future relationships and successes.

The Designer's Role in Value Engineering

The designer is an essential contributor to almost all VE proposals. A proposal should not be seen by the designer as a threat to the integrity of his or her original plans, but as an opportunity to participate in an innovative way with the partnering team.

When a VEJP is presented to the team, the designer should be present and prepared to act quickly in whatever capacity is required. The designer may be called upon by the team to make a quick initial evaluation of the proposal and to assess the feasibility of the idea. Often team members will rely on the designer to make his or her own contribution to the synergy of the situation.

There will be occasions when an idea cannot be implemented as originally submitted. Often the designer will see modifications that render it possible based on his more detailed knowledge of the plans. Most VEJPs would not be possible without active input from the designers.

A question that often arises is whether or not the designer should share in any savings generated from the value engineering proposal. In the true spirit of partnering all team members should benefit equally from their roles in the VEJP process. However a new set of problems arise where public agencies are involved. Foremost is the issue of the designer's contractual duty to provide an innovative set of plans in the first place. Thus he or she should not benefit from someone else's improvement to those plans. Others who are cynical about the engineering profession would suggest that the designer would be tempted to leave an innovation out of his plans in order to cash in on the benefits at some later date. The author does not agree with either of these assertions, but this perception is a major stumbling block to full implementation of the benefits of value engineering to all of the members of the partnering team.

While the author recognizes there is a need to share the benefits of a VEJP with all of the contract parties, it is clear that the public sector is unable do so across the board. The hope is that some day designers will be able to share in these VEJP savings just like the other partnering team members.

Value Engineering Examples

Many examples of value engineering could be shared with the reader in order to illustrate the issues discussed in this chapter. A few will be detailed in this section to illuminate the various points important to the concept of value engineering in the partnering environment.

On one project in northern Arizona, the contractor agreed to build several structures on a busy highway while maintaining through vehicular traffic. The original plans called for a phased construction with traffic running very close to the actual work being performed. In the partnering workshop, the idea of a temporary detour was suggested—to build a detour that would provide a bridge site clear of passing traffic. The idea would also improve the safety aspects of the project itself. The team decided that this idea had merit and assignments were given at the partnering workshop. Rather than use the state's designer for the detour layout, it was determined by the team that the contractor's engineer doing the construction surveying was in the best position to engineer the detour. This VEJP involved many additions and deletions to the contract with the net benefit to the project of about $20,000. This sum was split between the contractor and the state. The benefits included a four-month savings on the construction schedule for the structures on this project.

On this same project, the contract called for the contractor to build erosion control measures referred to in Arizona as "Rail Bank Protection." The contractor knew the state had stockpiled a large quantity of gabion baskets which had been deleted from another project earlier in the year. Consequently a substitution was proposed. The idea was accepted by the partnering team and implemented with a split of $60,000 in savings.

Another example, the Kitt Peak project near Tucson, required the contractor to construct a two-span bridge structure along with many other related items of work. The original contract called for the removal of approximately 100,000 cubic yards of material from the site with disposal off the project site. The contractor and Resident Engineer's staff were able to strike a compromise with the local Indian tribe, which relaxed the requirement to remove the material from the site. Instead the tribe allowed the contractor to simply recontour the slopes around the bridge. The ultimate value engineering savings on this $1,100,000 contract was $100,000 which was split between the state and the contractor.

There are many other examples which could be cited to illustrate the power of this synergistic relationship that has become known as partnering. Suffice it to say that value engineering is an important part of the partnering process and allows owners and contractors to be more profitable and to excel together in a successful project. As of this writing, the Arizona Department of Transportation has shared nearly $1,000,000 in value engineering savings with its partners on a wide variety of construction projects. Those who do not embrace this visionary approach to value engineering will ultimately deny themselves the significant savings that proposals offer to the industry.

Chapter 9

Joint Evaluation

The partnering process has its fundamental roots in the concept of Total Quality Management(TQM). It allows the owner and contractor to focus on this very important and profitable relationship. Those involved in the quality movement understand that the concept of continuous improvement is an essential element of TQM. Consequently, it should come as no surprise that the concept of continuous improvement is also very important to the partnering process.

Principles of Evaluation

The concept of continuous improvement requires some mechanism or process to assess current performance and then determine how that performance can be improved upon. Performance standards must be established and then evaluated on a regular basis. Plans must be adopted to further improve performance.

In Chapter 4, the development of the team charter was covered. This charter is the product of the team's efforts to define their common goals and objectives. Chapter 4 covered how the team must endeavor to draft their goals and objectives in a measurable way. This would allow the team to assess performance in an objective way on most or all of their goals and objectives. The evaluation segment of the partnering process will rely on these measurable goals and objectives to assess the progress of the team and determine ways that it can be improved.

A proper evaluation must rely on open and candid input from all of the team members. It is an opportunity to truthfully address the areas where team performance has been good and areas where the need for improvement is more obvious. This openness and honesty can come only from a team with a high level of internal trust and confidence in itself. Team members must understand that the input they receive is for the benefit of the team as a whole and that it is not intended as a personal affront to anyone. This level of trust is not developed overnight. It begins with the partnering workshop and continues to grow and mature through the course of the project.

Another important principle of the evaluation process is the involvement of all the team members. Each member of the team has a specific role on the project. All members are significant or they would not be a part of the team.

Therefore their input into team performance is important because of their unique perspective. If only managers filled out evaluations, then the team would bypass an important element of the project. They could possibly overlook areas where team performance was not meeting the original expectations.

The Joint Evaluation

The mechanics of the joint evaluation are rather simple. As part of the partnering workshop, team members must develop a process for team evaluation. This usually results in the production of a form that team members will fill out at specific intervals during the life of a project. Regardless of the process or tool used to evaluate the project, each team must feel comfortable with its own process and be satisfied that it will give them the feedback they need to effect continuous improvement.

Most joint evaluation forms are developed from the team's goals and objectives. If a goal or objective is important enough to appear in the partnering charter, then it probably is significant enough to be evaluated on a regular basis. A team may take these goals and objectives and modify or combine them to produce a succinct listing of the elements to be evaluated. Once this listing is agreed upon, the team should design the form or instrument that will be used to do the actual evaluations.

The team must determine what method of scoring will be used in these joint evaluations. A 1–5 or 1–10 system is popular and rather simple to implement. In this case each element on the evaluation is given a particular score by the team members. This allows them to then tally the responses and review the results from a numerical perspective. Progress can then be evaluated from an objective standpoint based on the scores given over a period of time.

Many teams choose to include an opportunity for members to provide narrative input as part of the evaluation process. This allows an individual to comment on an issue or problem that may not be obvious from the scores given on any specific goal or objective. Other teams require that if a score is below a certain level, then the person doing the evaluation must include some narrative comment. This provides the team with insight into the problem and the ability to take immediate action to solve the problem.

Any number of formats will work well for the joint evaluation. A sample is provided at Appendix E. Experience has shown that teams should develop forms which are simple to complete and which provide meaningful data. Teams should develop their own unique format and not be tempted to adopt someone else's form to save time.

The joint evaluation must be seen as an essential part of the partnering process and should not be underestimated in its significance. It can provide a team with early indications of problems starting to occur and allows team members an opportunity to resolve these issues early in the process. It gives the team an opportunity to improve the partnering process and refine the attributes that ultimately lead to success.

Chapter 10

Rewards and Recognition

Prior to the implementation of partnering, the owner and contractor were generally facing months and perhaps years of distasteful claims to resolve at the close of a project. The end of a construction project was not usually a cause for celebration. However on a partnered project, the team members find that there is much to celebrate in the achievements of the partnering team.

The Close-Out Workshop

Before the project team moves on to another project, it is advisable to have a short close-out partnering workshop to bring the team full circle in its relationship. This allows the project team the opportunity to celebrate its successes and learn from areas where problems occurred. It represents a closing of the loop on numerous issues for the team members and will serve to reinforce the partnering process on future projects.

The close-out workshop can be accomplished in two to four hours. Use of the original facilitator is helpful but not always necessary. In the absence of a facilitator, the team can use one or two of the key leaders from the project to run this important session. A typical agenda includes time allotted to a review of each of the team's goals and objectives. Also time should be allocated to reflect on the things that went well on the project in addition to the areas where improvement would have been possible. Given the distinct attributes of each project and team, the close-out workshop should be tailored to accommodate the specific needs of each.

Rewards and Recognition

Harkening back to partnering's tie to TQM, it should not be surprising that rewards and recognition are a part of the close-out workshop. The close-out workshop provides an excellent opportunity for the recognition of individuals on the partnering team for their specific efforts toward the overall success of the project. If appropriate, the CEOs can present "partnering champion" awards or plaques to one or several individuals who truly exemplified the principles of partnering on the project. There also may be cause to recognize a segment of the team that made a particularly significant contribution to the overall success of the project. The value of this type of recognition is great as reinforcement for future projects.

The issue of rewards and recognition is sometimes more difficult to deal with in the public sector. Many public entities function under procurement codes that prohibit meaningful rewards for their employees. These must be dealt with on a case-by-case basis. CEOs should provide whatever rewards are possible given the constraints of their individual organizations.

Assessing Success

An evaluation or assessment of the team's progress should be a part of the close-out workshop. This is accomplished by reviewing the team's measurable goals and objectives and making an honest evaluation of its achievements. The value of ensuring the measurability of these goals and objectives in the development of the team charter will be evident as the team seeks to assess their success on the project.

Prior to the close-out workshop, it is advisable that the team members be asked to make one final rating of the team and its performance against the charter. These evaluations can then be compiled and used as the basis for discussion of team performance at the close-out workshop. The purpose of this segment of the workshop is to allow team members the opportunity to frankly discuss the outcomes of the project team. During the many close-out workshops held by ADOT, it is not uncommon for team members to have many ideas for improving the process. Team members will also give credit to one another for their roles in the success of the project. The ultimate result is that the partnering team moves forward to the next project with a greater resolve for even more significant successes.

The rewards and recognition elements of the partnering process will not only recognize the achievements of the partnering team, but also reinforce the positive attributes of the process. It is an opportunity that ought not to be missed for further improving the process leading to greater benefits for all.

Chapter 11

Lessons Learned

Earlier in this book, the concept known as Total Quality Management (TQM) and how it relates to partnering was briefly covered. It was pointed out that partnering was a tool that addressed both the supplier and customer components of TQM. A fundamental element of any quality initiative is the need for continuous improvement. Over the last two years, the author has been involved in over one hundred partnered projects at ADOT. In addition, he has assisted many other organizations throughout the nation in their implementation of partnering. This experience has allowed him to identify a number of areas where improvements can be made and the process further refined. So in this spirit of continuous improvement, the author would like to share some of these lessons so the reader may take full advantage of this experience and benefit in the process.

Lesson Learned No. 1 Use outside facilitation at your partnering workshops until the owner and industry are mature in the process.
In Chapter 3 the need for facilitation was treated in some detail. The facilitator is important in the partnering process because he or she helps the team through the initial team building process. The facilitator also assists the team in developing the charter and mission of the partnering team. A good facilitator will unobtrusively make the partnering workshop flow. In fact, the best facilitators are so skilled in their role the teams hardly notice their presence during the workshop.

In the course of the last two years ADOT primarily has used outside facilitation in its partnering workshops. This was done for a couple of reasons. First, in the initial days of partnering, there still existed a lack of trust and bitterness between ADOT and the industry. Consequently, there was a real need to have a neutral third party present to facilitate the elements of the workshop. This helped everyone present put aside any past paradigms and move forward into the more productive and synergistic team relationships that partnering produces.

The second reason for outside facilitation was the basic lack of resources from within. For the first 20 or 30 workshops, the staff at ADOT was new to the partnering process and there were not many individuals who could have filled the facilitator's role. Thus it was determined that skilled facilitators would bring this expertise to the workshop.

Today there are some projects on which ADOT does not use outside facilitation. These are generally smaller projects or ones where the contractor's staff and ADOT's staff have previously worked together. On these projects there is usually an individual on the project staff with the skills necessary to facilitate and they fill this role for the team. The effectiveness of these individuals is dependent upon their skills and not necessarily on whether they have an engineering degree, represent management or have some specific role on the project. By the same token there are also individuals on staff with contractors who possess the skills necessary to facilitate a partnering workshop.

If an organization is just starting out in the partnering process then the advice the author would give is to use outside facilitation until you feel comfortable with the process. The movement to internal facilitation should be done in concert with the owner and the industry so that all parties are comfortable with the shift in approach. Not all facilitators should come from the owners staff but that over time facilitation in mature partnering relationships should be done by a broad range of individuals representing all partners.

Lesson Learned No. 2   The partnering workshop is the beginning of the process and should not be seen as the whole of partnering.

An important element of the partnering process is the partnering workshop. It provides the foundation for the partnering relationship and sets the stage for team success. Unfortunately, some fail to take full advantage of the power of this element of partnering.

Several areas contribute to the success of the partnering workshop. As was covered earlier, it is important to get all key team members to attend the workshop. Where this has not occurred the teams have been handicapped and frustrated in trying to develop the synergy necessary for success. Partnering leaders should endeavor to get suppliers, subcontractors and others to the workshop in addition to the more traditional attendees from the contractor's, designer's and owner's staffs.

Some who express disappointment with the partnering process have in fact held a partnering workshop. However, when queried about any follow-up they indicate that they have done nothing else to nurture and cultivate the partnering relationship. They think that once they have had the partnering workshop they have now partnered and the process is over. It is important for all to understand that the workshop is only part of the process and it alone does not constitute partnering.

Lesson Learned No. 3: No project over 12 months in duration should go without having a mid-project workshop.

The mid-project workshop was covered in detail in Chapter 7. It is intended to rejuvenate the stagnant partnering team and to charge the batteries of the participants where necessary. Experience has shown that many project teams fail to hold a mid-project workshop because they are too busy and do not have time. They admit that they continue to have problems, but they just do not have time to stop work to hold another workshop.

Unfortunately many partnering teams, while successful in their own right, are held back due to their reluctance to come back to the table and renew their relationship.

In order to address this problem, ADOT has implemented a policy that requires the team to gather together at some mid-point in the project if the duration of the construction is in excess of one year. While this has not been universally followed, there is clear evidence of the value of this reaffirmation of the team's goals and objectives and recommitment to the success of the project.

Lesson Learned No. 4: Take time to develop and apply the issue escalation process.

Where partnering relationships have been unsuccessful, there has generally been the lack of a well-defined issue escalation process. The ability for all parties to render timely decisions is fundamental to the success of a project. The issue escalation process, as described in Chapter 5, is the key to making this happen in the partnering relationship.

During one of ADOT's early partnering endeavors, the team evaluations came in after the first month with several comments which reflected a misunderstanding in the issue escalation process. Already decisions had been made that were causing problems on the project and the team was stalling in its efforts. Fortunately, both ADOT and the contractor recognized the problem as a breakdown in the team's understanding of the issue escalation process. With clarification, the problem was resolved and the team went on to experience success.

Lesson Learned No. 5: Develop and use the team evaluation process with consistency and commitment.

During the partnering workshop, the team should develop the evaluation form that will be used to monitor the efforts of all participants. This is important because of the need for continuous improvement within the team relationship on a particular project.

ADOT personnel have learned that it is important to develop a comfortable format for the team members which is truly a product of the team. A temptation for those experienced in partnering is to take a form from a previous project and force-fit it to the new project. Ultimately the fit will not be right, and the value of the evaluation will be less than satisfactory.

In developing the evaluation process, the team should err on the side of simplicity as opposed to a complex evaluation process. A simple numerical rating with an area for narrative comments is certainly sufficient for most projects. A technique that many teams have found valuable is to set a threshold in their scoring system that will trigger narrative comments. If for example the team has a scoring system of 1 to 10, it may decide that any rating below a 7 must have an accompanying narrative comment in order for the team to fully act on the problem this rating reflects.

Another element of the evaluation process concerns the review of the team evaluation forms. In many organizations where there are only a few projects, the CEO or someone in upper management has typically reviewed the monthly evaluations. During the initial stages of partnering implementation at ADOT, it was decided the State Construction Engineer would complete this review. This was done to ensure uniformity in the application of this new process and provided management with a first-hand view of the effectiveness of the evaluation process. It also allowed for minor course corrections on certain projects as problems became evident during the time of the project. As the number of projects increased, it became obvious that the State Construction Engineer would not be able to fulfill this role forever. With an average of 30 team members per project and 40 projects under construction, the State Construction Engineer was now looking at some 1,200 separate evaluations per month. Logically this was soon delegated to the districts for their use and action.

The team evaluation process can be a valuable tool in the TQM toolbox and will contribute to the improvement of the partnering team and its ultimate success.

Lesson Learned No. 6: Use all of the partnering tools on every project.
In the course of implementing partnering, some people have not been happy with the process. They are disappointed with the results and do not seem to be receiving the same benefits that other owners and contractors are clearly enjoying. Often when the author queries these individuals, he finds that they are not really partnering or are only partially partnering. Some will have had only a workshop. Others will not have developed a charter. Many that do not experi-

ence success have not developed an issue escalation process or do not evaluate their team's effectiveness on a regular basis. The lesson to be learned here is, in order for the process to work, all of the elements of partnering must be used.

Lesson Learned No. 7: Decide early on whether record keeping will be done and then measure only those areas necessary.

The issue of record keeping will come up in any organization involved in significant partnering efforts. Ironically partnering advocates echo some of the same arguments as do many TQM leaders when discussing measurement and record keeping. For example some insist that measurement is essential to quality. Others say that the measurement process detracts from the more fundamental and significant elements of TQM. Ultimately each organization will have to reconcile this issue in line with its own set of needs and circumstances.

ADOT determined early on that it would measure several key indicators in order to track the impacts of the partnering process. These indicators included value engineering, timeliness of completion, claim trends and field overhead costs. It was felt that these indicators could be readily measured without a significant amount of administrative activity and that they would offer evidence of the effectiveness of ADOT's partnering program.

The first step toward meaningful measurement is to determine what the pre-partnering condition is so that you have a baseline to measure from. Having made the measurement decision early, this was not a major problem at ADOT. A lesson ADOT learned in this area is that it will be harder to determine the baseline data if you try to collect it late in the implementation process.

Another record keeping issue concerns the reasons for measuring. An organization needs to understand why it is keeping records about partnering so it can determine which areas to measure. Some organizations want to use partnering to reduce claims or litigation. Still others want to improve relations with industry. Regardless of the motivation, a key lesson learned is that measurement activities must be tailored to suit the organization's needs.

Lesson Learned No. 8: Top management commitment ultimately determines if an organization is successful and profitable in the partnering process.

The most important ingredient for success in partnering is the commitment of the CEO and other members of top management. Too many issues in partnering require organizational changes that can only be implemented with unquestioned CEO support. Without these changes partnering will be used in a limited way with corresponding limited success.

This topic has been discussed in detail in Chapter 4 so the author will not dwell on it here. Having been one of the pioneers in partnering, ADOT has been involved with hundreds of companies, agencies and trade organizations as they have embraced the partnering concept. Almost without exception, those who struggle with the implementation of partnering can trace their problems back to CEOs who are not committed to the concept. They may say the right things and nod their heads at the appropriate times, but if it does not come from their heart then it will ultimately impact their outward behavior in support of their field staff.

Other Lessons Learned

In addition to the previous listing of important lessons learned, there are a number of do's and dont's that should be considered by any organization contemplating a partnering endeavor. A few of these are provided as follows:

Do's

1. Study and learn about the partnering process before trying to implement it on a project.
2. Talk to owners, contractors, trade associations and anyone else who might give you insights into the partnering process.
3. Spend the time getting top management focused on the partnering process and prepared for the cultural changes necessary for success.
4. Spend time as an industry learning together about partnering so that everyone has the same understanding of the process.
5. Always invest the time necessary for planning the partnering workshop.
6. Always use all of the partnering tools.
7. Require that the escalation process be actively used on the project.

Dont's

1. Don't partner if the CEO and top management are not absolutely committed to the process.
2. Don't assume that just because partnering is so logical that everyone will embrace it without question.
3. Don't short-change the time spent in the partnering workshop.
4. Don't stop partnering after the workshop.
5. Don't assume that everything is going okay unless you have the monthly evaluations to back it up.
6. Don't forget to nurture the partnership.
7. Don't forget to reward and recognize your partnering champions.

While the author has cited just a few of the lessons that have been learned across the nation over the last couple of years, many others could have been described. Each organization should track their own lessons and make improvements to the partnering process within their own operations. This refining process will provide powerful results to those committed to the concept of continuous improvement.

Chapter 12

Results of Partnering

In the early days of partnering, no tangible results could be pointed out as evidence of its benefits. The true pioneers of partnering implemented it because they had vision and knew it was the right thing to do. They did it because they had faith that partnering represented a better way to do business. To the nay sayers and doubters they said "trust us."

After two years of partnering at the Arizona Department of Transportation there is no longer a need to say "trust us." The results are in and offer tangible evidence of the benefits partnering can produce.

Claims

ADOT was originally motivated to introduce partnering due to a significant problem with construction claims. Figure 1 reflects the claim trends at ADOT prior to partnering. The number of claims and their value rose year after year in spite of all the traditional efforts by ADOT to stem the tide. Measures such as new specifications, claims seminars and more detailed project documentation were just some of these efforts.

While the payout on claims was high, a greater toll was exacted in resources dedicated to efforts that brought no tangible improvements to Arizona's transportation system. Perhaps more significant was the personal price paid by all those involved in ADOT projects over the years. ADOT anticipated that partnering would greatly impact the history of claims in Arizona.

The results of partnering speak for themselves. During the first two years of partnering in Arizona, there has not been a single new claim or issue which has not been resolved through the issue escalation process. ADOT changed from an organization where great effort went into claims resolution to one where it no longer needed to invest manpower in this way. At some point in the future an issue may not be resolved through the escalation process. If this occurs, then it will be turned over to an Alternate Dispute Resolution (ADR) process. A single claim once in a while is far better than the multitude of claims that made up the pre-partnering condition.

An unanticipated result in the claim area resulted when agency and contractor employees alike started implementing the principles of partnering on non-partnered projects. ADOT found that you cannot work synergistically on one

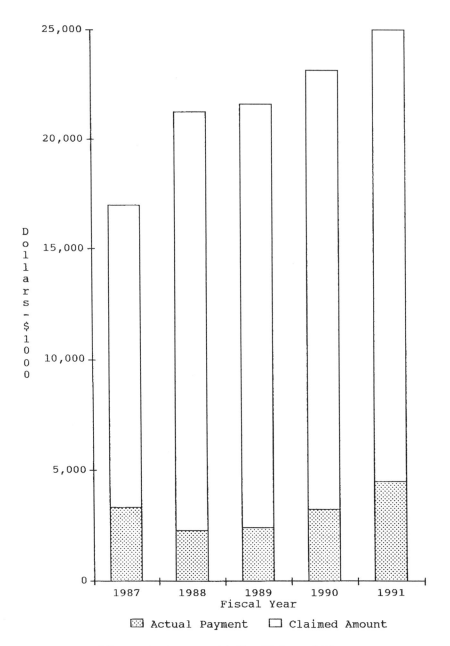

Figure 1. Average Dollar Value of Claims

project then drive down the road and revert to the old style of doing business on an adjacent project. Consequently there was a clear trend on the non-partnered projects where ADOT and contractors resolved many long-standing issues by applying the principles learned through the partnering process.

It is obvious that the claim scenario in Arizona has been significantly changed due to partnering. As agencies experience the benefits of partnering, they need to examine their previous resource allocation to the administration of claims. With fewer claims ADOT is in the process of examining its budget for legal counsel. In Fiscal 1994 ADOT's budget for legal counsel was reduced by $134,000 with an additional reduction of $80,000 anticipated in Fiscal 1995. In addition, the in-house staff for claims administration had been reduced. There now appears to be an opportunity to divert scarce public resources to higher-value activities such as additional transportation facilities for the traveling public.

Construction Time

Prior to partnering, timely completion of construction projects was a significant problem in Arizona. In Fiscal 1991, 27% of all projects failed to be completed in the original contract time. The direct cost impacts to ADOT, contractors and most importantly the public, were thought to be quantified by the liquidated damages assessed under the terms of the contract. However, the liquidated damages did not come close to the value of the indirect and unquantifiable costs that all parties incurred when contracts were not finished on time.

While the original motivation to partner at ADOT came from the serious claim problem, contract time has also surfaced as a very real benefit. The trend on completed projects reflects this point. To date construction time on completed partnered projects has been reduced by over 20%. No longer is one in four projects completed after the original end of contract time: rather the question now is how much ahead of schedule the project will be finished. Figure 2 shows a scatter plot of completion times for those projects completed to date.

What is it about the partnering process that promotes this reduction in construction time? Certainly the fact that the owner's representatives and the contractor's staff are now working *together* toward the common goal of project completion has something to do with it. Energies are now focused on the completion of the project as opposed to impeding the other party from completing their part of the contract. Partnering teams work together to help other team members be more effective and efficient in the performance of their duties. Taken together the elements of partnering contribute to the substantial benefits associated with timely project completion.

One example of this phenomenon is the Elliot Road Traffic Interchange project in Phoenix, which was scheduled to be completed in 13.5 months and opened in 8.5. Another is the Rose Garden Traffic Interchange where contract time was set for a 16.5-month duration and the project was completed in 8. It has been gratifying to see these and other benefits of parties working together on the partnering team.

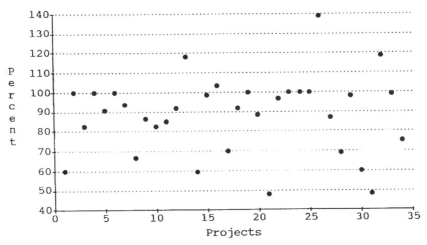

Figure 2. Contract Time Used versus Completed Partnering Projects

When projects are completed early, who benefits? The contractor benefits by earning his profit in a shorter period of time, by incurring less project overhead and by increased opportunity for resource utilization. The owner benefits through reduced construction costs, lower contract administration costs and a better product. However, the real winner is the pubic or the end user of the project. They receive the project ahead of schedule and are able to take advantage of its uses earlier than anticipated. The public is the real winner when the partnering process is applied in the public sector. Of course in the private sector, greater profitability for all parties to the contract heads the list of benefits accrued.

Construction Administration

As projects are completed ahead of schedule, claims eliminated and procedures improved, it is logical to anticipate a reduction in the construction administration/engineering expenses of the owner. Depending on the owner's organization, these engineering and administration costs can range from 5% to 15% or more. The dollar value of this activity also varies from state to state with ADOT's annual expenditure in the $35,000,000 range.

Two factors contribute to the final total of engineering costs on a project. The first factor is the expense of the Resident Engineer and his staff and the second is the expense of the designer in his post-design role during construction. ADOT has seen a substantial reduction in the costs associated with the field staff component of its construction administration costs through the elimination of low-value activities. The second has actually gone up slightly due to the activities detailed in Chapter 6. However, the net benefit to ADOT, when both components are summed, is a 24% reduction in construction administration costs.

Value Engineering

The role of value engineering in the partnering environment has been covered in Chapter 8. The partnering team has the necessary synergy to make value engineering a powerful tool. In the two years since ADOT started partnering in Arizona it has saved approximately $1,000,000 which has been shared with contractors. As partnering proliferates the concept of value engineering will continue to be an important tool in the partnering tool chest

Budget Reduction

As owners plan their construction programs, it is necessary to consider many components of a project's total cost. The following relationship is generally accepted as a reflection of total cost:

$$\begin{array}{r} \text{Contractors Bid} \\ +/- \text{ Supplemental Agreements} \\ - \text{ Value Engineering} \\ + \text{ Claims} \\ \hline \text{Total Project Cost} \end{array}$$

At ADOT the difference between the Total Project Cost and the contractors bid was historically an increase of about 5%. This was the contingency value added to each contract at the time of award and became the new project budget.

Partnered projects completed to date have provided further evidence of additional savings. Comparing the Total Project Costs for completed partnered projects reveals that the contingency value is only 3% instead of the 5% that has been the historical trend in Arizona. This savings of 2% of the total construction budget represents significant funds. These funds can now be programmed for additional unscheduled projects or to move up the completion timetable on others. Of particular note is the fact that the components making up the Total Project Cost are now changed under the partnering scenario. For example, with no claims, the aggregate value of supplemental agreements is less and there is an increase in value engineering savings. The net effect is that ADOT need only program 3–4% contingency rather than the standard 5% that was used for so many years. Certainly this is a tangible manifestation of the benefits of partnering.

Indirect Benefits

While many real benefits can be attributed to partnering, some are less tangible. For example, as projects are completed early the liability exposure of both the owner and the contractor is reduced. This cannot be measured or quantified on a project-by-project basis, but it is a very real issue in the daily business practice of both parties.

For years, contractors have stated that they will adjust their bids upward depending upon who the owner's representative is going to be on a particular project. Today the question is "how much will the bid be reduced if the contractors know that this will be a partnered project? How can organizations quantify the benefits of such a relationship? We know they exist even though it is difficult to assign a specific numerical value. The evidence is that partnering impacts the bottom line in any organization.

Putting Partnering Savings to Work

Partnering savings can be real, but if there is not a mechanism to measure them on a routine basis, they will evaporate and leave nothing more than a good feeling. After a few months of feeling euphoric about all the money ADOT was saving, it decided a process was needed to capture these savings for future use.

Working with the staff of the Administrative Services Division, ADOT was able to establish an accounting system that monitors the savings being generated on partnered projects. This system will allow ADOT to accumulate the partnering savings in specific accounts until there is enough money available to fund a project that either is added to the construction program or advanced to an earlier year.

Partnering benefits are real. They speak for themselves. While they may vary in amount from project to project, the trends are clear and unmistakable. Will there ever be a project that won't experience these benefits? Certainly. Partnering cannot overcome all of the possible plans problems or personalities that may exist in the world today. But think of how badly the project would have gone without partnering.

Not all organizations are keeping records of their partnering savings. It certainly is not a requirement for success. Each owner, designer and contractor must satisfy its own specific needs for keeping records and measuring success. If the results in Arizona seem too good to be true, the reader is referred to such states as Florida, Texas, Minnesota and others where partnering is actively applied and similar results are being experienced. After literally hundreds of partnered projects across the country the advocates of partnering no longer need to say, "trust me." The many benefits speak for themselves.

Chapter 13

Strategic Partnering

The focus of preceding chapters of this book has been the use of partnering on a project-by-project basis. While this is a powerful use of the partnering concept, it is possible for to apply partnering even more completely within an organization. In essence the application of partnering on specific projects represents a tactical approach to suppler relationships. In this chapter the concept of partnering will be addressed from a strategic point of view.

Project vs. Global Solutions

During the workshop process there is an opportunity to resolve project-specific problems which the team has identified. If approached in the spirit of partnering, this investment of time will pay handsomely in saved time and effort down the road during actual project construction. Typical problems resolved on partnered projects include the engineering shop drawing review process, materials testing and inspection procedures, the acceptance of materials manufactured off the project site and documentation for the purpose of making payment to the contractor. Each team came up with its own unique solution for dealing with these problems so progress was not impacted.

After a number of partnered projects, ADOT noted that issues began to repeat themselves with some regularity with each team solving its own version of the problem. It became clear that these common problems should be solved as ADOT or agency issues instead of each project team working independently. If such problems were solved once and for all at the agency level, the teams could then focus their attentions on more specific project-level issues. This was the beginning of one phase of ADOT's strategic partnering initiative.

Recently this particular phase of strategic partnering has been applied to a large interchange project in Phoenix. It includes a large amount of structure work and the reconstruction of major portions of Interstate 10 and US 60. It is complex from a traffic-volume standpoint with numerous phases required to accomplish the work and yet maintain traffic flow. Before the partnering workshop, a special meeting was held with top management from all of the organizations connected with the project. The purpose of this meeting was to educate these leaders in the concept of strategic partnering.

During the partnering workshop, the team was empowered to go beyond routine partnering activities. They addressed issues from the perspective of resolving them for all of the other construction projects in the state. If they resolved the problem of timely review of engineering shop drawings, then that solution would be applied to the internal processes within ADOT. Special groups were established within the partnering team to work on specific problems that were both project and agency related. Representatives from ADOT, the contractor and the subcontractor participated on these teams.

As of this writing, this project is still underway, and it is too early to pass judgment on this phase of ADOT's strategic partnering initiative. However, this particular team is in a position to make a substantive improvement in the overall business practice of ADOT's highway construction program. In the spirit of continuous improvement, this expanded approach to partnering can only enhance its application on individual projects.

Strategic Partnering with Suppliers

In 1991 Charles E. Cowan was appointed as the Director of the Arizona Department of Transportation. He quickly learned of the poor relationship that existed between the agency and contractors and designers. Many problems clearly could be attributed to this relationship with these key suppliers of services to ADOT.

Before he could address this issue, it was necessary to fully understand both the magnitude of the problem and its many nuances. Consequently Mr. Cowan commissioned a survey of contractors and engineering designers. It consisted of a broad range of questions focused on the theme of "What are we (ADOT) doing that is costing you time and money?"

The survey was sent out in 1991 to 126 suppliers and over 100 ADOT employees. In addition 31 personal interviews were conducted. The response rate to the mail-out surveys was a phenomenal 80% which gives you some idea of the interest level in this subject on the part of the contracting and design communities in Arizona. A consulting firm was retained to conduct the interviews and compile the results of the survey. This was done to ensure the anonymity of the respondents and to encourage an open and honest assessment on their part. The responses to the survey were frank and in many cases brutally honest. While there were literally hundreds of comments, three very specific themes came through loud and clear. They are as follows:

- ADOT needed to do something about making fair and timely decisions.
- ADOT needed to establish problem-solving procedures.
- ADOT needed to address the issue of fair and timely payment to contractors.

In addition to the narrative comments returned, the respondents were asked to score various areas of ADOT's performance. On a scale of 1 to 5, those surveyed rated such areas as relations, timeliness of decisions and overall effectiveness of ADOT. The tabulated results of some of these areas are found in Figures 3, 4 and 5.

A number of interesting observations can be made. The low scores given by contractors in every area becomes obvious from the ratings. Also contractors gave consistently lower ratings than everyone else in all categories. In addition, ADOT staff, whether in the field or design offices, rated themselves better than their industry counterparts. Clearly the problems experienced by industry were more severe than was realized in viewing them from an internal perspective.

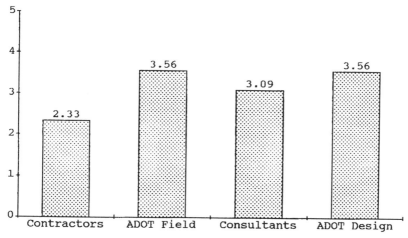

Figure 3. Overall Effectiveness of ADOT—1991 Supplier Survey

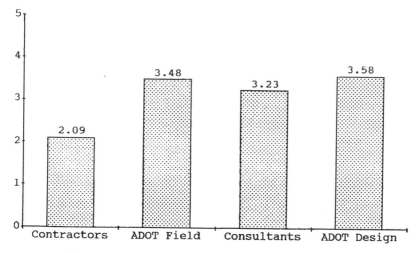

Figure 4. Relations—1991 Supplier Survey

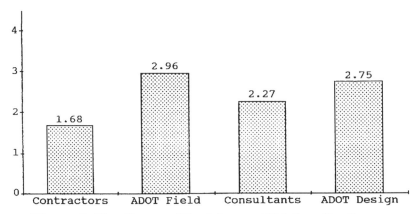

Figure 5. Timeliness of Decisions—1991 Supplier Survey

It is one thing to solicit input from suppliers, but it is a totally different matter to take action on that input. ADOT determined that these same suppliers needed to participate not only in the problem identification but also in the resolution. Consequently in January 1992, ADOT gathered about 200 individuals from both industry and within the organization to review the survey results and identify specific action items. By the end of the day 143 issues were compiled and ADOT/industry teams were established to address each one.

During the course of the next year these teams met and worked on their assigned problems. A progress report was presented to ADOT management and leaders of industry on a quarterly basis in order to establish accountability for the improvements and activities of this strategic partnering effort.

At the end of the year, most of the 143 issues had been resolved and solutions put in place. However proof of resolution could only be confirmed with industry validation. So in the fall of 1992, another survey patterned after the first effort was conducted. This time there was a total of 150 mail out surveys and 35 interviews. The response rate for the mailed surveys was only 51% this time. Figures 6 through 8 reflect the data compiled from the two surveys and show the changes in the responses between the two years.

Several items about the 1992 results are worth noting. First, the scores were higher than the first survey in almost every category. Second, the largest changes in score were from the contracting community. Some nearly doubled over the previous survey. Another observation that can be made from the second survey is that the ADOT design staff rated all three areas lower than in 1991. A full explanation of this situation is difficult to arrive at due to a major change in work load and a reorganization that occurred in this area during the survey period. While the marked improvement was gratifying, it is also clear that improvement can continue to be made.

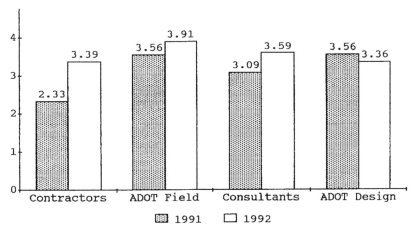

Figure 6. Overall Effectiveness of ADOT—
Comparison of 1991 and 1992 Surveys

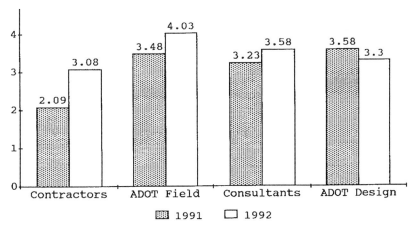

Figure 7. Relations—Comparison of 1991 and 1992 Surveys

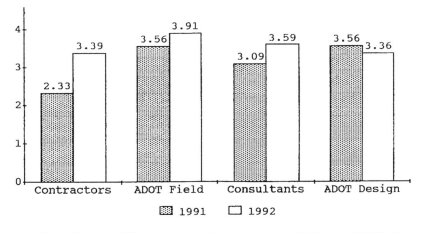

Figure 8. Timeliness of Decisions—Comparison of 1991 and 1992 Surveys

The process of involving the industry in resolving issues raised in the survey was again repeated. Representatives from all elements of the contracting and engineering design communities were invited to attend another one day workshop in February 1993. New issues were identified and teams were organized. These teams are now working on their problems on the same basis as they did in the first round of strategic partnering.

Where does strategic partnering go from here? In the fall of 1993 there will be another supplier survey sent out to assess progress and to identify additional areas for improvement. More teams will be established to address the new issues and further improvements will be made. This cycle of assessing, planning, implementing and validating will continue each year to further improve the business practice of the Arizona Department of Transportation. This is the spirit of TQM and continuous improvement that drives ADOT in its quest for quality.

Chapter 14

Non-Construction Applications of Partnering

The focus of this book has been the application of partnering to the construction project. It was the construction industry that was originally so fed up with its problems that it searched for a better way to do business. However the astute reader will have already noted that the basic principles and elements of partnering can be applied universally to nearly every relationship.

The benefits of partnering are clear. Greater productivity, more effective use of resources, reduced disputes, time reductions and dollar saving are the most evident. These benefits are not unique to the construction industry. They can be achieved in any supplier-customer relationship.

Applying the Principles of Partnering

The process of applying the principles of partnering is rather simple. First identify the stakeholders involved. Next determine what goals are shared by the stakeholders. Third establish a process for resolving disputes (i.e., an issue escalation ladder). Fourth, provide a way to evaluate the team and fifth, recognize and reward the achievements of the team.

Once an organization begins to implement partnering, it will find a host of other opportunities where partnering principles apply. ADOT chose to do internal partnering between sections that heretofore had not worked well together. Managers and supervisors will soon discover that they cannot treat their contractors with honesty, respect and integrity and then treat their employees otherwise. The principles of partnering should begin to permeate all relationships, both internal and external.

Non-Construction Examples

There are a number of examples of partnering applications outside of the construction arena. A few will be presented here to illustrate them.

In 1992 a major design effort was begun for I-10 in Tucson. A system of frontage roads were to be designed and built under difficult circumstances. The project involved multiple design firms, several companies, representatives from the community and ADOT staff. Over a two-day period these participants held a partnering workshop to determine how best to complete the design portion of this project. A charter was prepared and issues identified and addressed. The main difference between this effort and a construction project is the final product is a set of plans instead of a completed highway project.

Another example comes from the construction administration section in ADOT. For years there was a need for a comprehensive automation effort in this area. In January 1992 a partnering workshop was held which included ADOT construction staff, ADOT Information Services Group, the local vendor for the hardware and the manufacturer of the PC hardware. Again common goals were established, a charter was signed and issues resolved. It was the beginning of a very successful project. Not long ago this team got together when it neared the completion of one of its milestones and held a follow-up workshop. New issues were raised and resolved. The successes of the earlier efforts were recognized and celebrated. This project is destined to radically change the approach to automation efforts in the state of Arizona.

One of the of the major activities that ADOT is involved in is the collection of fees and taxes from motor carriers as they travel through the state. Not long ago it was decided that ADOT needed to improve its relationship with the motor carrier industry and resolve some of our common problems. A one-day workshop was held with 35 representatives from ADOT and industry where common goals were identified and a charter developed. Problems were listed and teams established to resolve these problems. These teams are now working to resolve issues that range from dealing with ports of entry to a very controversial taxation system to collect revenue for the state's highway fund. While not every problem went away at the end of the day, this workshop does mark the beginning of a new relationship with this important group of customers and suppliers.

Another example of a non-construction application of partnering occurred as a precursor to the 1994 legislative session for the state of Arizona. Over 50 individuals from all segments of the transportation industry participated in a first-ever partnering session to jointly prepare legislation that would be proposed to the lawmakers. ADOT, the motor carrier industry, legislators, contractors and private citizens all had an opportunity to present their issues and hear input from those others present. Knowing that among the common goals of this group is to have a world-class transportation system that treats all participants fairly has caused a unification that has never been present before. To be sure, there will still be differences among these participants. However, the intended product of this workshop is to move forward with joint legislation founded on common goals that will ultimately benefit all the people who live and travel in the state.

The list of non-construction applications of partnering is long. Many organizations are using the partnering process in nearly every relationship in which they are involved. The opportunities are limited only of the vision of the individual or organization applying the principles.

Chapter 15

A Vision of Partnering

The concept of partnering is exciting, rewarding and destined to forever change the construction industry of the United States. Already it has netted substantial savings to the Arizona Department of Transportation and the people of Arizona.

Partnering represents an opportunity for owners, designers, contractors, sub-contractors and suppliers to maximize their individual abilities in a synergistic arena. This is not to say that some among this group have not previously entered into relationships akin to the high-performing partnering teams. For example in the private sector, some developers, large corporations and others have worked closely with selected contractors and suppliers because they recognized the value of such a relationships. These efforts certainly set the stage for formalized partnering relationships to occur in the public sector and the low-bid, fixed-price environment.

Today partnering has taken the country by storm. In just over two years, literally hundreds of organizations have embraced the concept and applied its principles. A recent survey by the FMI Corporation, indicated that 89% of all state departments of transportation have used partnering/team building to some degree or another on their construction projects.[1] Cities, towns, counties and school districts are finding partnering a valuable tool that stretches their scarce capital improvement dollars. Professional and trade organizations across the country have endorsed the partnering concept as the bold and innovative process that it is. Clearly this is more than a passing fad or whim.

The vision of partnering goes far beyond ADOT and Arizona. When one considers the savings that other departments of transportation are and will generate through the use of this management concept, the magnitude is staggering. Scarce public and private resources will no longer be wasted on needless claims and litigation. Projects will consistently be delivered on or ahead of schedule. Those who use our nation's transportation systems will experience a quantum improvement in the services they receive. Transportation systems in the United States would become world-class again.

With this vision of partnering, one could consider the savings to be generated if all public and private construction contracts were partnered. This impact on the economy of Arizona and the rest of the country would be phenomenal. Our nation's infrastructure will vastly improve as money and other resources are focused on high-value activities.

Then there is the vision of partnering as it relates to all other supplier and customer relationships. As an effective tool of Total Quality Management, partnering will cause these relationships to flourish and become more effective. Profits will soar, customer service improvements will be dramatic and the United States will reemerge as the economic leader it once was.

This is the vision of the future of partnering.

Reference

1. Highway Construction Industry Survey on Project Partnering. Page 1. FMI Corporation. Raleigh. NC 27612.

Appendix A

Partnering Specification
Arizona Department of Transportation

Section 104.0(B)

The Arizona Department of Transportation intends to encourage the foundation of a cohesive partnership with the contractor and its principal subcontractors and suppliers. This partnership will be structured to draw on the strengths of each organization to identify and achieve reciprocal goals. The objectives are effective and efficient contract performance and completion within budget, on schedule, and in accordance with plans and specifications.

This partnership will be bilateral in make-up, and participation will be totally voluntary. Any cost associated with effectuating this partnering will be agreed to by both parties and will be shared equally.

To implement this partnering initiative prior to starting of work in accordance with the requirements of Subsection 108.02 and prior to the preconstruction conference, the contractor's management personnel and ADOT's District Engineer will initiate a partnering development seminar/team building workshop. Project personnel working with the assistance of Construction Operations Services will make arrangements to determine attendees at the workshop, agenda of the workshop, duration, and location. Persons required to be in attendance will be the ADOT Construction Supervisor and key project personnel. The contractor's on-site project manager and key project supervision personnel of both the prime and principal subcontractors and suppliers. The project design engineers. FHWA and key local government personnel will also be invited to attend as necessary. The contractors and ADOT will also be required to have Regional/District and Corporate/State level managers on the project team.

Follow-up workshops may be held periodically throughout the duration of the contract as agreed by the contractor and the Arizona Department of Transportation.

The establishment of a partnership charter on a project will not change the legal relationship of the parties to the contract nor relieve either party from any of the terms of the contract.

Appendix B

Partnering Special Provision
Texas Department of Transportation

"Partnering"—The Department invites the Contractor to join it in a voluntary "partnering" arrangement for the work covered by this contract. Acceptance of the partnering invitation may result in a no-cost time extension, and a delayed notice to proceed until the team-building workshop and the pre-construction conference have been held. If the partnering invitation is accepted:

1. Contractor may select and provide a third-party facilitator to conduct the team building workshop for the Contractor and Department personnel. Facilitator selection shall require Department concurrence. An anticipated cost of approximately $3,000 to $7,000 for the facilitator and his associated expense will be shared equally by the Department and the Contractor. The Contractor will be reimbursed by the Department for the Department's portion on the first monthly estimate on an extra work basis.

2. Contractor and Department will exchange lists of the key personnel to be participants in the workshop. The list will contain the name and job title of each person, a contact phone number and the address for job related correspondence. The Department will furnish each of the identified personnel with instructions, travel reservations and materials for the workshop.

3. The Contractor and the Department will be responsible for any expense incurred by their respective employees, including meals, travel and lodging.

4. A definitive working arrangement for the partnership will be agreed upon and committed to in writing at the workshop. The arrangement will set out the mutually recognized goals and expectation of each of the parties.

5. Contractor and Department agree that each of the key personnel identified will be assigned to the work for its duration and will not be transferred or reassigned without 30 days notice and adequate replacement.

6. Contractor and Department agree to provide at the work site persons who will be committed toward the achievement of the goals and implementation of the partnership agreement.

7. All disputes will be processed in the manner agreed upon by the parties during the orientation.

8. Follow-up workshops may be held periodically throughout the duration of the contract as agreed by the Contractor and the Department.

9. Either partner may withdraw from the Partnership Arrangement upon written notice to the other. However, no claim or dispute settled or change approved during the existence of the partnership shall be revived.

10. The sole remedy for non-performance of the partnership shall be termination of the Partnership Arrangement as set out in paragraph 9 of this section.

Appendix C

Covenant of Good Faith and Fair Dealing, Arizona Department of Transportation Section 104.01(A)

This contract imposes an obligation of good faith and fair dealing in its performance and enforcement.

The contractor and the Department, with a positive commitment to honesty and integrity, agree to the following mutual duties:

a. Each will function within the laws and statutes applicable to their duties and responsibilities.
b. Each will assist in the other's performance.
c. Each will avoid hindering the other's performance.
d. Each will proceed to fulfill its obligations diligently.
e. Each will cooperate in the common endeavor of the contract.

Appendix D

Sample Partnering Charter

PARTNERING AGREEMENT

We, the partners involved in the I-10/Elliot Road Interchange Project, commit to trust, cooperation and mutual respect in providing a quality, timely, and safe project within budget.

To accomplish this, we commit to the following objectives:

- Build a quality project, building it right the first time
- Utilize Value Engineering in joint cooperation to achieve goals
- Limit cost growth to two-percent
- No claims or litigation
- No lost time accidents/minimize third party claims
- Complete the project in seven months
- Ninety-five percent of decision-making at the project level
- Demonstrate commitment to the issue resolution process
- Cost of project within owner and contractor budget, with the contractor profiting
- Be proactive in designing a joint public awareness program and to minimize local business disruptions
- Enhance mutual trust between partners
- Finalize the project within sixty days of acceptance
- Limit the final inspection to one preliminary punch list and one final punch list
- Streamline paperwork, change orders and administrative procedures
- Explore avenues for using contractor quality control samples as an adjunct to ADOT's acceptance
- Have fun!

81

Appendix E

Joint Evaluation

Gila River Bridge Partnering Team
Joint Evaluation

Date:_____

Goal or Objective	Rating
Communications	_____
Administration	_____
Safety	_____
Project Progress	_____
Community Relations	_____
Issue Escalation	_____
Joint Evaluation	_____
Quality	_____
Claims	_____
Willingness to Meet Commitments	_____
Health of the Partnership	_____
Total	_____

Comments:_____
_____
_____
_____
_____
_____

Note: Goals should be scored between 1 (poor) and 10
(excellent). If a score of 7 or less is given then comments
are required to explain the rating.

Name:_____ Signature:_____

Index